The HUMAN MIND EXPLAINED

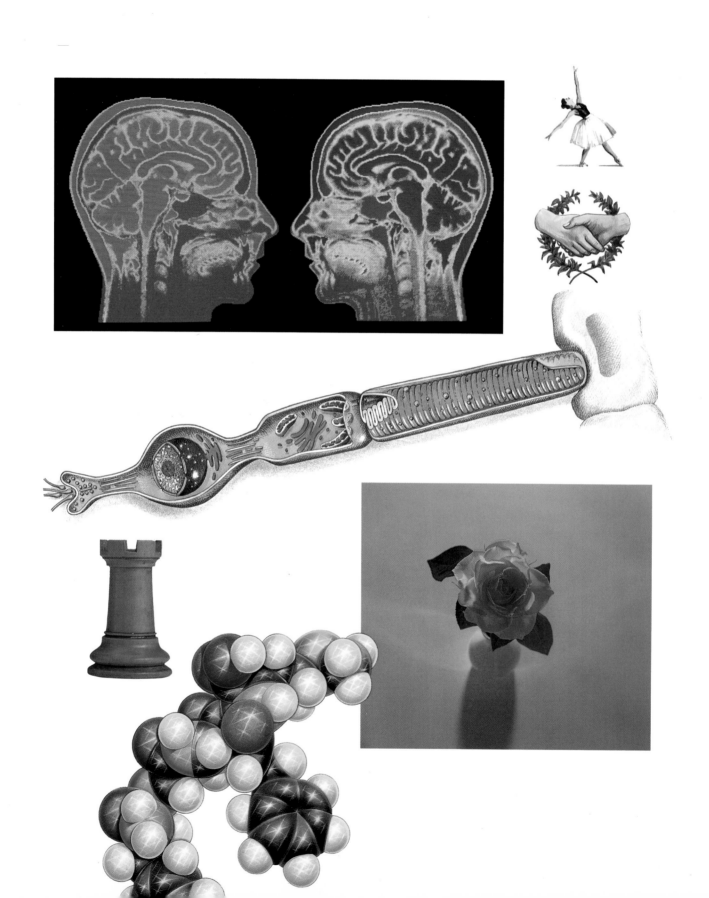

The HUMAN MIND EXPLAINED

An Owner's Guide to the Mysteries of the Mind

Susan A. Greenfield, General Editor

A HENRY HOLT REFERENCE BOOK

HENRY HOLT AND COMPANY

NEW YORK

Contents

A Henry Holt Reference Book
Henry Holt and Company, Inc.
Publishers since 1866
115 West 18th Street
New York, New York 10011

Henry Holt ® is a registered trademark of
Henry Holt and Company, Inc.

Copyright © 1996 by Marshall Editions
Developments Limited
All rights reserved

**Library of Congress
Cataloging-in-Publication Data**

The human mind explained: an owner's guide
to the mysteries of the mind / Susan Greenfield,
general editor. — 1st ed.
   p.   cm. — (A Henry Holt reference book)
Includes bibliographical references and index.
   1. Intellect. 2. Brain.
   3. Psychology, Comparative.
I. Greenfield, Susan. II. Series.
BF431.H748 1996   96–6139
153—dc20   CIP
ISBN 0-8050-4499-X

Henry Holt books are available for special
promotions and premiums. For details
contact: Director, Special Markets.

First American Edition—1996

Conceived, edited, and designed by
Marshall Editions, London

Printed and bound in Italy by New Interlitho Spa
Originated by HBM Print, Singapore

All first editions are printed on acid-free paper.∞

10 9 8 7 6 5 4 3 2 1

Project editor	Jon Kirkwood
Art editor	Simon Adamczewski
Picture editor	Zilda Tandy
Research	Jolika Feszt
	Michaela Moher
Design assistant	Eileen Batterbury
Assistant editor	Isabella Raeburn
Copy editor	Maggi McCormick
DTP editors	Mary Pickles
	Pennie Jelliff
	Kate Waghorn
Managing editor	Lindsay McTeague
Editorial director	Sophie Collins
Index	Caroline S. Sheard
Production	Sarah Hinks
Contributors	Jerome Burne
	John Farndon
	Steve Parker

Previous page (clockwise from top):
magnetic resonance gives an outside view;
having ideas about physical and
interpersonal skills; seeing the light about a
retinal rod; a rose inspires emotional states;
brain chemicals that stop you from feeling
pain; having more ideas about chess.

Overleaf (clockwise from top):
where the brain keeps track of the sense of
touch; facial nerves show how the body
links to the brain; one of the brain's cells;
bigger brains helped the evolving mind;
color vision depends on contrasting views.

Foreword 7

Introduction 8

Surveying the Mind

Introduction 11

What is a brain? 12

Discovering the brain 14

The working mind 16

Comparing brains 18

Inside the mind 20

Electronic minds 22

Discovering through damage 24

Probing the mind 26

The view from outside 28

Brain map 1 30

Brain map 2 32

Map of functions 34

Map of chemicals 36

Building the Brain

Introduction 39

Building blocks 40

Support and protection 42

The brain's cells 44

The electric cell 46

To fire or not to fire? 48

Sending signals 50

Crossing the gap 52

The gap gallery 54

On or off 56

Unlocking the gate 58

Discovering transmitters 60

Remaking the mind 62

Recovering from damage 64

Food for thought 66

The brain plan 68

The developing brain 70

The cell factory 72

Linking up 74

The aging brain 76

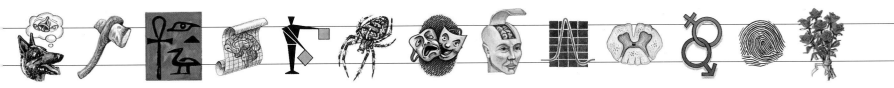

Inputs and Ouputs

Introduction 79

Body links 80

The long junction 82

On the move 84

Guided motion 86

Seeing the light 88

Contrasting views 90

Levels of seeing 92

Active vision 94

Sensing sound 96

The mind's ear 98

Sense of touch 100

Pain pathways 102

Feeling pain 104

A matter of taste 106

On the scent 108

The primal sense 110

Staying upright 112

The big picture 114

Keeping control 116

Dealing with drives 118

Rhythms of the mind 120

Survival sense 122

Far Horizons

Introduction 125

The evolving mind 126

Conditioning the mind 128

The infinite store? 130

Active memory 132

Meeting of minds 134

Word power 136

Parallel minds 138

Having ideas 140

First thoughts 142

Learning to be human 144

Milestones of the mind 146

Male and female 148

Thinking together 150

States of Mind

Introduction 153

Emotional states 154

The unique self 156

Being aware 158

The resting mind? 160

Abnormal states 162

Talking cures 164

Physical cures 166

Feeling low 168

The mind adrift 170

Anxious states 172

Fears and fixations 174

The failing mind 176

Drugs of abuse 178

Motivation 180

The moral mind 182

Individual minds 184

Bibliography 186

In short 186

Measure for measure 186

Index 187

Acknowledgments 192

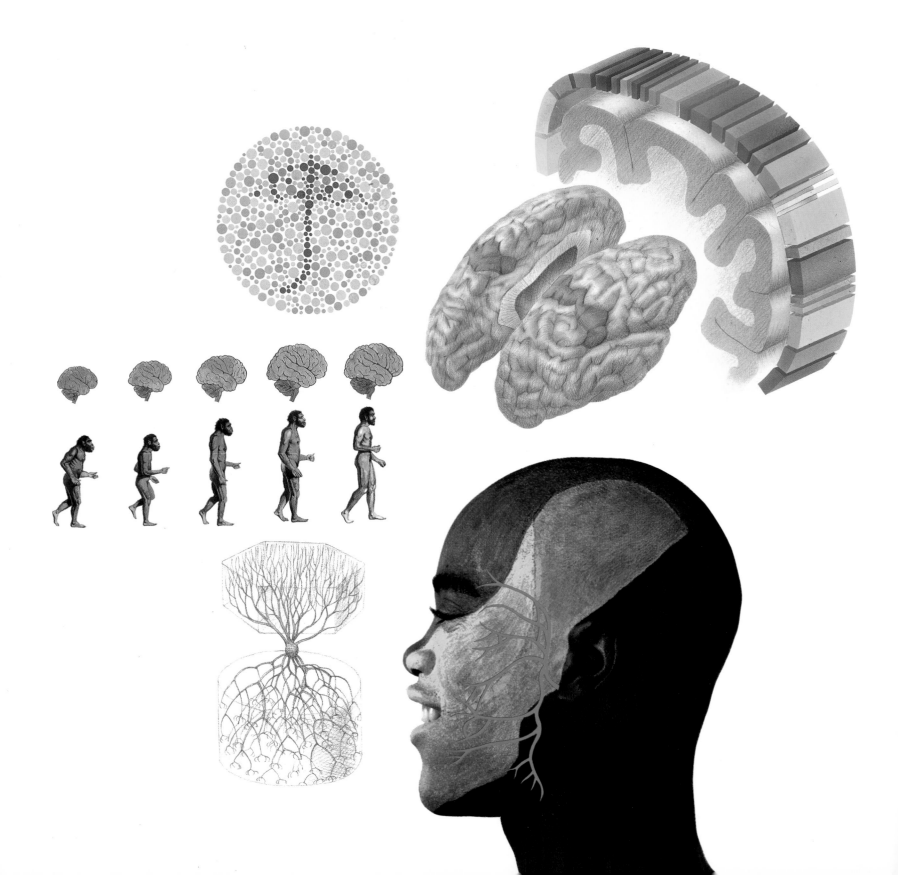

Foreword

The human brain presents the ultimate riddle: how can a mass of tissue with the consistency of raw egg be responsible for your "mind," your thoughts, your personality, your memories and feelings, and even your actual consciousness? Until recently, this was still a conundrum, but the last 25 years have witnessed the most astonishing progress in facts, clues, and ideas.

An organ like any other in the body, the brain is nonetheless tantalizingly different. Whereas transplants of hearts, kidneys, and lungs are now well established and widely accepted, the brain remains the essence of the individual. All our experiences of the outside world are dependent on the workings of this creamy mass of cells as numerous as the trees in the Amazon rainforest, and all crammed into the skull. The brain processes information from the senses about sound, light, touch, taste, and smell, and can initiate action, coordinating movements ranging from running and climbing to whispering and piano playing. These powers enable us to interact with a changing environment and to convey our thoughts and ideas to others; without these abilities we would each be a prisoner in our own private world.

Over the years scientists and thinkers have tried to fathom the mysteries of the mind. Here are some of the techniques they have used, along with observations of patients with different types of brain malfunction. We see what is known about how the brain works, how it processes information, and how it controls the body. We look at thoughts, emotions, and memories, and how they are modified by drugs, therapeutic and otherwise, and by other types of clinical treatments. Empowered by our incredible mind, we explore what drives us, constrains us, and inspires us.

Susan A. Greenfield

Introduction

This book's five interlinked sections, each looking at different aspects of the mind, reveal the brain's workings and structure, how we interact with the world around us, how we interpret and make use of what we experience, and the internal world of thoughts and feelings. In this way, the knowledge of how the brain is put together and how it performs its tasks is made accessible and understandable.

SURVEYING THE MIND

Starting with an inquiry into the nature and workings of the mind, **Surveying the Mind** moves on to the most powerful and up-to-date techniques for investigating the brain and then provides vital anatomical and functional maps.

BUILDING THE BRAIN

The myriad microscopic components of the brain – its cells – are the key to understanding how it all works. **Building the Brain** explores their structure and function, and sees how they change from conception to old age.

INPUTS AND OUTPUTS

Sensory signals stream into the brain, keeping it in touch with what is going on outside and inside the body. **Inputs and Outputs** shows how these signals are made and interpreted, and how the brain sends out its messages of control.

FAR HORIZONS

Learning, memory, and the growth from infancy to adulthood of a new human mind are extraordinary enough, but **Far Horizons** goes a step farther and probes the world of the intellect and the higher functions of the brain.

STATES OF MIND

We know that we are self-aware and that life can have its emotional ups and downs, yet it is hard to grasp just what makes up each unique human mind. **States of Mind** looks at how we experience the positive and negative aspects of being alive and explores the boundaries, if any, to thought and behavior.

It is relatively simple merely to describe the components and systems of the human mind, but much more exciting to understand how and why they work as they do. Using analogies with familiar items and events, and with the help of straightforward language, this book provides explanations of a kind not easily attainable otherwise.

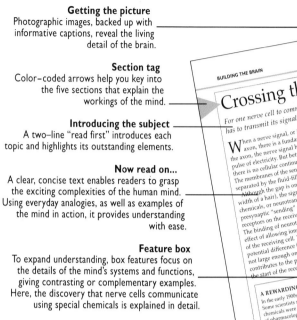

Getting the picture
Photographic images, backed up with informative captions, reveal the living detail of the brain.

Section tag
Color–coded arrows help you key into the five sections that explain the workings of the mind.

Introducing the subject
A two–line "read first" introduces each topic and highlights its outstanding elements.

Now read on...
A clear, concise text enables readers to grasp the exciting complexities of the human mind. Using everyday analogies, as well as examples of the mind in action, it provides understanding with ease.

Feature box
To expand understanding, box features focus on the details of the mind's systems and functions, giving contrasting or complementary examples. Here, the discovery that nerve cells communicate using special chemicals is explained in detail.

HOW TO USE THIS BOOK

Photographs and illustrations are the starting points that lead to knowledge and an understanding of the nature and workings of the mind. Words link closely to pictures to explain in depth the science of the human brain. Each of the book's double-page spreads is a self-contained story. But since topics do not always fit neatly and completely under the headings superimposed on them, connections lead from the edge of each right-hand page to other topics, both within the same section and in different sections.

In order to promote flexibility of thought and depth of interest, the connections made are often deliberately wide-ranging. The connections can be used to make the book interactive and help forge links of understanding between different, yet complementary, aspects of how the brain works and the mind thinks.

Three sections – **Surveying the Mind, Building the Brain,** and **Inputs and Outputs** – are devoted to exploring and explaining specific areas of the brain's function and performance. The other two – **Far Horizons** and **States of Mind** – deal more widely with the processes of thought and the way we experience our mind in action.

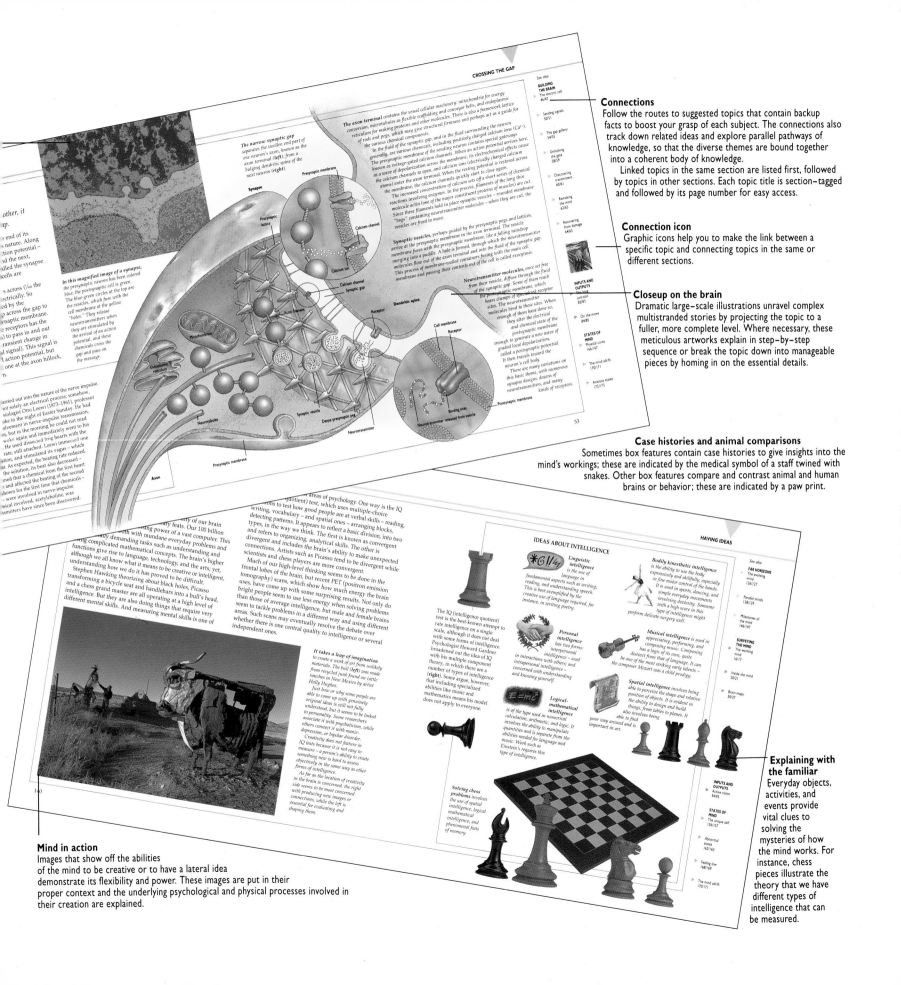

The axon terminal contains the usual cellular machinery: mitochondria for energy conversion, microtubules as flexible scaffolding and conveyor belts, and endoplasmic reticulum for making proteins and other molecules. There is also a framework lattice of rods and pegs, which may give structural firmness and perhaps act as a guide for the various chemical components.

In the fluid of the synaptic gap, and in the fluid surrounding the neuron generally, are various chemicals, including positively charged calcium ions (Ca²⁺). The presynaptic membrane of the sending neuron contains special gateways known as voltage-gated calcium channels. When an action potential arrives here, as a wave of depolarization across the membrane, its electrochemical effects cause the calcium channels to open, and calcium ions (electrically charged) start across the membrane, the calcium channels quickly start to close again.

The increased concentration of calcium sets off a short series of chemical reactions involving enzymes. In the process, filaments of the long thin molecule actin (one of the major constituent proteins of muscles) are cut. Since these filaments hold in place synaptic vesicles – rounded membrane "bags" containing neurotransmitter molecules – when they are cut, the vesicles are freed to move.

Synaptic vesicles, perhaps guided by the presynaptic pegs and lattices, arrive at the presynaptic membrane in the axon terminal. The vesicle membrane fuses with the presynaptic membrane, like a falling raindrop merging into a puddle. A hole is formed, through which the neurotransmitter molecules flow out of the axon terminal and into the fluid of the synaptic gap. This process of membrane-walled containers fusing with the main cell membrane and passing their contents out of the cell is called exocytosis.

Neurotransmitter molecules, once set free from their vesicle, diffuse through the fluid of the synaptic gap. Some of them reach the postsynaptic membrane, which bears clumps of specialized receptor sites. The neurotransmitter molecules bind to these sites. So, enough of them have done so, they alter the electrical and chemical state of the postsynaptic membrane enough to generate a new wave of graded local depolarization, called a postsynaptic potential. It then travels toward the neuron's cell body.

There are many variations on this basic theme, with numerous synapse designs, dozens of neurotransmitters, and many kinds of receptors.

The narrow synaptic gap separates the swollen end part of one neuron's axon, known as the axon terminal (left), from a bulging dendritic spine of the next neuron (right).

In this magnified image of a synapse, the presynaptic neuron has been colored blue; the postsynaptic cell is green. The blue-green circles at the top are the vesicles, which fuse with the cell membrane at the yellow "holes." They release neurotransmitters when they are stimulated by the arrival of an action potential, and these chemicals cross the gap and pass on the message.

Labels on illustration: Presynaptic membrane · Synapse · Presynaptic lattice · Calcium channel · Calcium ion · Calcium channel · Synaptic gap · Dendritic spine · Receptor · Cell membrane · Receptor · Actin filament · Mitochondrion · Endoplasmic reticulum · Binding sites · Microtubules · Neurotubules · Postsynaptic membrane · Synaptic vesicle · Dense presynaptic peg · Neurotransmitter · Neurotransmitter released from vesicle · Presynaptic membrane · Axon

See also

BUILDING THE BRAIN
The electric cell 46/47

Sending signals 50/51

The gap gallery 54/55

Unlocking the gate 58/59

Discovering transmitters 60/61

Remaking the mind 62/63

Recovering from damage 64/65

INPUTS AND OUTPUTS
Tactile junction 82/83

On the wing 84/85

STATES OF MIND
Physical cures 166/167

The mind adrift 170/171

Anxious states 172/173

53

(left-hand column, lower continuation)

...carried out into the nature of the nerve impulse, not solely an electrical process; somehow, physiologist Otto Loewi (1873–1961), professor ...woke in the night of Easter Sunday. He had ...involvement in nerve-impulse transmission, ...but in the morning he could not read ...He woke again and immediately went to his ...He used dissected frog hearts with the ...rate, still attached. Loewi immersed one ...tion, and stimulated its vagus – which ...the solution, the beating rate reduced. As expected, its beat also decreased – ...oned that a chemical from the first heart ...and affected the beating of the second ...shown for the first time that chemicals – ...were involved in nerve-impulse ...chemical involved, acetylcholine, was ...transmitters have since been discovered.

Connections
Follow the routes to suggested topics that contain backup facts to boost your grasp of each subject. The connections also track down related ideas and explore parallel pathways of knowledge, so that the diverse themes are bound together into a coherent body of knowledge.
Linked topics in the same section are listed first, followed by topics in other sections. Each topic title is section-tagged and followed by its page number for easy access.

Connection icon
Graphic icons help you to make the link between a specific topic and connecting topics in the same or different sections.

Closeup on the brain
Dramatic large-scale illustrations unravel complex multistranded stories by projecting the topic to a fuller, more complete level. Where necessary, these meticulous artworks explain in step-by-step sequence or break the topic down into manageable pieces by homing in on the essential details.

Case histories and animal comparisons
Sometimes box features contain case histories to give insights into the mind's workings; these are indicated by the medical symbol of a staff twined with snakes. Other box features compare and contrast animal and human brains or behavior; these are indicated by a paw print.

(lower left text columns)

...ity of our brain ...ssing power of a vast computer. This ...th with mundane everyday problems and ...ally demanding tasks such as understanding and ...functions give rise to language, technology, and the arts; yet, although we all know what it means to be creative or intelligent, understanding how we do it has proved to be difficult.
Stephen Hawking theorizing about black holes, Picasso transforming a bicycle seat and handlebars into a bull's head, and a chess grand master are all operating at a high level of intelligence. But they are also doing things that require very different mental skills. And measuring mental skills is one of

It takes a leap of imagination to create a work of art from unlikely materials. The bull (left) was made from recycled junk found on cattle ranches in New Mexico by artist Holly Hughes.
Just how or why some people are able to come up with genuinely original ideas is still not fully understood, but it seems to be linked to personality. Some researchers associate it with psychoticism, while others connect it with manic-depression, or bipolar disorder.
Creativity does not feature in IQ tests because it is not easy to measure – a person's ability to create something new is hard to assess objectively in the same way as other forms of intelligence.
As far as the location of creativity in the brain is concerned, the right side seems to be most concerned with producing new images and connections, while the left is essential for evaluating and shaping them.

...areas of psychology. One way is the IQ ...quotient) test, which uses multiple-choice ...ons to test how good people are at verbal skills – reading, writing, vocabulary – and spatial ones – arranging blocks, detecting patterns. It appears to reflect a basic division, into two types, in the way we think. The first is known as convergent and refers to organizing, analytical skills. The other is divergent and includes the brain's ability to make unexpected connections. Artists such as Picasso tend to be divergent while scientists and chess players are more convergent.
Much of our high-level thinking seems to be done in the frontal lobes of the brain, but recent PET (positron emission tomography) scans, which show how much energy the brain uses, have come up with some surprising results. Not only do bright people seem to use less energy when solving problems than those of average intelligence, but male and female brains seem to tackle problems in a different way and using different areas. Such scans may eventually resolve the debate over whether there is one central quality to intelligence or several independent ones.

IDEAS ABOUT INTELLIGENCE

Linguistic intelligence is the use of language in fundamental aspects such as writing, reading, and understanding speech. This is best exemplified by the creative use of language required, for instance, in writing poetry.

Personal intelligence has two forms: interpersonal intelligence – used in interactions with others; and intrapersonal intelligence – concerned with understanding and knowing yourself.

Logical-mathematical intelligence is of the type used in numerical calculation, arithmetic, and logic. It involves the ability to manipulate quantities and is separate from the abilities needed for language and music. Work such as Einstein's requires this type of intelligence.

Bodily kinesthetic intelligence is the ability to use the body expressively and skillfully, especially in fine motor control of the hands. It is used in sports, dancing, and simple everyday movements involving dexterity. Someone with a high score in this type of intelligence might perform delicate surgery well.

Musical intelligence is used in appreciating, performing, and composing music. Composing has a logic of its own, quite distinct from that of language. It can be one of the most striking early talents – the composer Mozart was a child prodigy.

Spatial intelligence involves being able to perceive the shape and relative position of objects. It is evident in the ability to design and build things, from tables to planes. It also involves being able to find your way around and is important in art.

The IQ (intelligence quotient) test is the best-known attempt to rate intelligence on a single scale, although it does not deal with some forms of intelligence. Psychologist Howard Gardner broadened out the idea of IQ with his multiple component theory, in which there are a number of types of intelligence (right). Some argue, however, that including specialized abilities like music and mathematics means his model does not apply to everyone.

Solving chess problems involves the use of spatial intelligence, logical-mathematical intelligence, and phenomenal feats of memory.

See also

FAR HORIZONS
The evolving mind 126/127

Parallel minds 138/139

Milestones of the mind 146/147

SURVEYING THE MIND
The working mind 16/17

Inside the mind 20/21

Brain maps 30/37

INPUTS AND OUTPUTS
Active vision 94/95

STATES OF MIND
The unique self 156/157

Abnormal states 162/163

Feeling low 168/169

The mind adrift 170/171

Mind in action
Images that show off the abilities of the mind to be creative or to have a lateral idea demonstrate its flexibility and power. These images are put in their proper context and the underlying psychological and physical processes involved in their creation are explained.

Explaining with the familiar
Everyday objects, activities, and events provide vital clues to solving the mysteries of how the mind works. For instance, chess pieces illustrate the theory that we have different types of intelligence that can be measured.

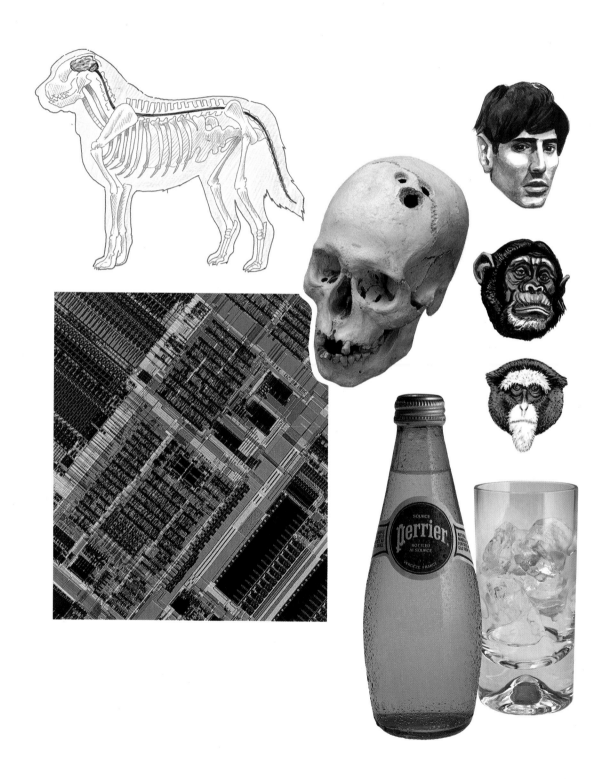

Surveying the Mind

Millennia ago, at about the time when people started attempting to explore and explain the world around them, they also began to consider thought itself. Generation after generation of great thinkers came up with theories about what the brain did and how the mind worked – a process that continues today, for we seem irresistibly drawn to investigate ourselves.

We are unique among species in that as well as performing everyday functions, we are able to study these actions and our behavior. We also examine thought processes and how the brain works. For many years, work in this field was often inaccurate and based largely on guesswork, but the advance of science has opened many avenues of inquiry, whether by experiment, comparison, or observation. Tools to examine the brain systematically have now become available, enabling us to probe its innermost workings and set out detailed maps of its anatomy and function.

*Left (**clockwise from top**): are bigger brains best?; thought processes in something as simple as getting a drink; how human and silicon brains measure up; the advancing animal brain; letting out spirits through a hole in the head.*
*This page (**top**): changing personality the hard way; (**right**) what animals can do that we cannot, and vice versa.*

What is a brain?

Why do we – and most other creatures living on Earth – have brains? And how did they evolve?

Inside the head of every person is one of the most astonishing objects in the universe: the human brain. It looks like a rather squishy overgrown walnut and is little bigger than two clenched fists. Yet it enables us to think, feel, and act – to read books and build televisions and to be human.

Humans are far from unique in possessing a brain – nearly every animal on Earth has one. What all brains have in common is that they enable creatures to respond to a situation – whether that situation is simply an obstacle in its path or the problem of finding food or fleeing from an enemy. Via the nerves, a brain both receives information (from the rest of the body and the outside world) and sends out orders (to control body functions and initiate action).

Without a brain to direct them, creatures that need to move and find food would find it hard to survive. The brain is thus a command and control center, in which the specialized cells that interpret and process data and then send out orders to the body are grouped together. Grouping makes the command and control process much more efficient than it would be if these cells were scattered around the body.

Single-celled organisms were the first to appear, perhaps 3.8 billion years ago. They still exist today, and while some of them, such as the amoeba, can move, they have no nerves. Next to evolve were multicelled organisms, such as sponges, again with no nerves. Specialized nerve cells first appeared in such creatures as jellyfish.

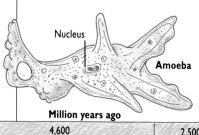

Nucleus

Amoeba

Million years ago

Segmented worm

Cerebral ganglion

Ventral nerve cord

A segmented worm, like its ancestors that evolved over 500 million years ago, has a cerebral ganglion – a cluster of nerve cells in its head – and a ventral nerve cord.

A brain is a device that enables an organism to control its response to the world. When a cat stalks its prey and then charges, its brain is being fed data concerning sight, hearing, smell, taste, and touch from its sense organs. The brain decides how to respond to this information and sends out appropriate commands to the body to crouch, move, and pounce.

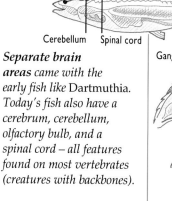

Olfactory bulb

Cerebrum

Fish

Cerebellum   Spinal cord

Insect

Cerebral ganglion

Nerve cord

Ganglion

Separate brain areas came with the early fish like Dartmuthia. Today's fish also have a cerebrum, cerebellum, olfactory bulb, and a spinal cord – all features found on most vertebrates (creatures with backbones).

Insects such as Collembola, like other invertebrates, have brains and nervous systems, but the structure is more varied than with vertebrates. Insects have more sense organs, for example, antennae and leg hairs.

Cerebellum
Cerebrum

Olfactory bulb

Reptile

In early reptiles, Hylonomus and Hypsognathus for example, brain structures grew larger, and distinct hemispheres developed.

Collembola   Hylonomus

Dartmuthia

4,600	2,500	570	500	440	410	365
Pre–Cambrian era	Pre–Cambrian era	Cambrian	Ordovician	Silurian	Devonian	Carboniferous

See also

SURVEYING THE MIND
▶ The working mind
16/17

▶ Comparing brains
18/19

▶ Brain maps
30/37

INPUTS AND OUTPUTS
▶ The big picture
114/115

▶ Dealing with drives
118/119

▶ Survival sense
122/123

FAR HORIZONS
▶ The evolving mind
126/127

THE CREATURE THAT EATS ITS BRAIN

Only an organism that moves from place to place requires a brain. An organism that stays still responds automatically to changes in its environment, but has no need to direct its movements. Plants often have sophisticated reactions – like turning their leaves to face the sun – but they do not have to move, so possess no brain.

The relationship between a brain and mobility is illustrated by a tiny marine creature, the sea squirt. It swims about like a tadpole when young. But when it matures, it attaches itself permanently to a rock and feeds by filtering plankton out of the water. It then consumes its own brain because it does not need it any more.

In the early part of its life, the sea squirt Ascidian *has a brain and nerve cord to control its movements (**right top**). As it reaches its sedentary mature form, these structures are gradually absorbed and digested, leaving only those needed for filter feeding (**right bottom**).*

In early mammals, like Megazostrodon, *the cerebrum developed dramatically, creating a frontal lobe on each hemisphere. In more recent mammals, the cerebral* hemispheres swelled enormously, taking over all but the most basic automatic responses, such as control of breathing and heartbeat.

In birds, like Archaeopteryx, there was more growth at the back of the brain.

Mammal

Cerebrum
Cerebellum
Spinal cord

Only animals with a head have true brains. Heads developed because animals move forward and have their sense organs at the front of the body. The brain processes the sensory input and issues commands.

Spinal cord

Bird

Cerebellum
Cerebrum
Spinal cord

Our brain and nervous system have much the same layout as those of all other vertebrates, from fish to dogs. At the heart of all vertebrate nervous systems is the central nervous system (CNS), which consists of the brain, including the cerebellum and cerebrum, and the spinal cord – the bundle of nerves running up to the brain through the backbone. All the other nerves in the body converge on the CNS. Some carry sensory data to the CNS; others carry commands from the CNS to the rest of the body.

The human brain is a product of gradual evolution. Like the brain of every other vertebrate, it has been created by adaptation to changing conditions over hundreds of millions of years.

Only in the last few million years of this evolution has it been the brain of a "hominid," a humanlike creature such as our ancestor Australopithecus afarensis. Before that, its ancestry can be traced back through apelike creatures such as Ramapithecus, primates, small mammals, reptiles, and fish. As the brain evolved, it grew outward in successive layers. Even today, its structure reveals its ancestry, with the most primitive feature – the brain stem – at the center, and the most modern – the cerebrum with its highly developed outer layer, the cortex – on the outside.

Human

Cerebrum
Cerebellum
Spinal cord

| 290 Permian | 245 Triassic | 210 Jurassic | 140 Cretaceous | 65 Tertiary | 2 Quaternary |

Hypsognathus

Megazostrodon

Archaeopteryx

Tyrannosaurus

Ramapithecus

Australopithecus afarensis

Discovering the brain

For millennia, great thinkers have applied their minds to the problem of how, and with what, we think.

Thinking and the brain are, to us, inextricably linked, but this link was not always so clear. One of the first to realize that thoughts occur in the brain was the Greek physician Hippocrates (c.460–377 B.C.), who believed that the mind was in the brain and controlled the body, but was something intangible. Unlike his contemporaries, he also believed that disease and head injuries, not demons, caused madness and affected coordination.

This split between mind and body, or between mind and brain, became known as mind–body dualism. The opposite argument, that the mind and brain or mind and body are one – that all thoughts are basically physical processes in the brain cells – is called monism. The dualism–monism debate has dominated thinking about the mind since the theories were first proposed.

We now know that all thought occurs in the brain and, indeed, where specific processes, such as perception or language, take place. Yet there is still no agreement as to where consciousness – our awareness of ourselves and our identity – fits in. Some argue that mind and brain are the same thing, that consciousness is a physical, mechanical entity and that one day we may be able to fashion a mechanical mind which experiences consciousness as we do. Others suggest that our minds and consciousness are very different from the mere mechanical activity of brain cells. Yet others steer a middle course, arguing that consciousness is a mechanical process of the whole brain, rather than individual brain cells.

VITALISM, OR THE ROMANTIC DILEMMA

Ever since it became clear that the brain was the seat of thought, many have resisted the idea that the mind is simply a machine. Antoni van Leeuwenhoek (1632–1723), who was the first to study nerve cells with a microscope, believed that the brain contained a special vital animal spirit or fluid that embodied our life force and consciousness. Members of the late 18th-century Romantic movement, led by German writer Johann Wolfgang von Goethe, argued that the brain contained an active life force – an animating energy that turned its tissue into something living and dynamic. Although this idea was long ago discredited by scientists, it is not so different from some modern ways of looking at consciousness.

From prehistoric times until quite recently, certain peoples carried out trepanning – the drilling of a hole in the skull. The theory behind this practice was that the sufferer's mind had been taken over by demons or evil spirits which had to be let out. But it may in fact have helped alleviate certain physical symptoms.

The ancient Greek philosopher Plato (428–348 B.C.) agreed with Hippocrates that the mind was in the head, but believed that the mind and body are separate, and that we reach the truth not via our senses but through our thoughts – by logic and reasoning.

Aristotle (384–322 B.C.), Plato's student and rival, thought the mind – and the seat of our feelings – was in the heart. He also argued that the mind and brain were one and that the mind is entirely physical, with the result that we can only understand the mind by studying the body.

René Descartes (1596–1650), the French philosopher, perpetuated Plato's mind–body dualism by insisting that the human body and the senses were entirely mechanical and material – but the mind was something else, and that the seat of this nonmaterial mind was the pineal gland.

The British philosopher John Locke (1632–1704) argued that mind and body were two aspects of the same unified phenomenon: the mind needs the body to gain experience through the senses and the body needs the mind to store and use this experience.

James Mill (1773–1836), the Scottish philosopher, argued in his book Analysis of the Phenomenon of the Human Mind *that the mind is merely a machine and can be explained entirely in physical terms.*

Immanuel Kant (1724–1804) tried to resolve the mind–body debate with the idea of faculties (mental powers) of sense, understanding, and reason which integrate the mind and body. This German philosopher also argued that we are born with some knowledge (a priori knowledge), but learn through experience (a posteriori knowledge).

The Greek physician Erisistratus (304–250 B.C.) divided the brain into two regions: big (the cerebrum) and small (the cerebellum). He also noticed how many more wrinkles the human brain had compared with those of animals and deduced, correctly, that they indicated greater brain power.

At much the same time, from his dissections made in Alexandria, the Greek anatomist Herophilus (born 320 B.C.) worked out that there were two kinds of nerves: sensory nerves, which received sense impressions; and motor nerves, which stimulated motion.

In the 1700s many people still believed that the nerves were tubes containing a mysterious vital fluid. Then the Swiss physiologist Albrecht von Haller (1708–77) showed they were simply the carriers of impulses that stimulated muscles or created a sensation and that all nerves lead to the brain and spinal cord – which must, therefore, be where the senses are perceived and responses initiated.

The German physiologist Johannes Müller (1801–58) took a crucial step in showing that the perception of the senses takes place in the brain when he demonstrated that each of the sensory organs responds to stimuli in its own way. For instance, if the optic nerve leading from the eye to the brain is stimulated, we see a flash of light, regardless of whether light was the stimulus.

Galen (A.D. 129–199), the Greco–Roman physician, is seen by many as the founder of experimental physiology.

He learned his trade by tending to gladiators' wounds. He believed cerebrospinal fluid was the psychic fluid, the fluid of the mind. When he discovered the ventricles in the brain where this fluid collects, he believed that he had found the seat of the mind.

The German physician Franz Gall (1758–1828) believed, rightly, that different parts of the brain are linked to different parts of the body. He also believed, wrongly, that personality and the relative development of particular regions of the brain could be read from the pattern of bumps on the skull, so initiating the pseudo-science of phrenology.

Frenchman Paul Broca (1824–80) was one of the first modern brain surgeons, treating an abscess on the brain, for instance, by trepanning (cutting through the skull). He also showed through post-mortems how damage to a certain area on the left frontal lobe impaired speech – establishing for the first time a link between a particular area of the brain and a particular ability.

The Flemish anatomist Andreas Vesalius (1514–64) refused to accept Galen's work in its entirety, but built on it through study and dissection. In addition, Vesalius finally discredited Aristotle's idea that the mind, and the seat of our feelings, was in the heart, insisting that it was in the brain. Such was his standing and the authority of his book on anatomy, De Humani Corporis Fabrica, first published in 1543, that no one has since doubted this assertion.

In this drawing of the brain which appeared in Vesalius's great work, the membranes have been stripped from the brain to disclose the cerebral hemispheres connected by the corpus callosum.

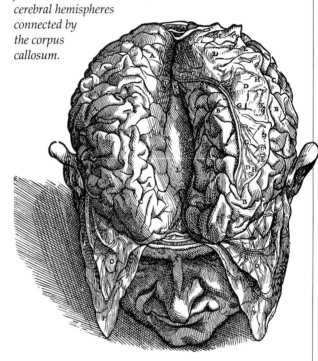

Hermann von Helmholtz (1821–94), the German scientist and philosopher, was a pupil of Johannes Müller at the Medical Institute in Berlin. Helmholtz was a pioneer in the study of both the eye and the ear, and invented both the ophthalmoscope, which is used to examine the interior of the eye through the pupil, and the ophthalmometer, with which faulty vision is measured, so allowing the correct lenses to be prescribed. He also showed that sensory conduction by the nerves (and thus the "speed of thought") was measurable. By demonstrating that the mind was thus limited by the physiology of the nerves, he opened the way to the experimental study of the mind.

See also

SURVEYING THE MIND
▶ What is a brain? 12/13

▶ The working mind 16/17

▶ Comparing brains 18/19

▶ Discovering through damage 24/25

▶ Probing the mind 26/27

▶ Brain maps 30/37

BUILDING THE BRAIN
▶ Building blocks 40/41

INPUTS AND OUTPUTS
▶ The big picture 114/115

FAR HORIZONS
▶ The evolving mind 126/127

▶ Parallel minds 138/139

STATES OF MIND
▶ Being aware 158/159

The working mind

Numerous theories of how the mind works have arisen from the drive to know more about ourselves.

For thousands of years, only the brain was studied by scientists; the mind was seen as the realm of philosophers. But by the mid-1800s, scientists had found that sensations, at least, were the result of nerve impulses that could be analyzed and measured. If this were so, might not science tackle perceptions, feelings, and memories as well as the brain itself? The German physiologist Wilhelm Wundt (1832–1920) believed it could and opened the first psychological laboratory at the University of Leipzig in 1879.

Wundt tried to study the mind by "introspection" – the analysis of one's own thoughts as they occur – but by the early 1900s, many psychologists seriously questioned this approach. The American John B. Watson (1878–1958), for instance, argued that with introspection results can never be proved or disproved. He maintained that a more valid approach was to study the way people behave and what they actually do and to analyze how people or animals respond to a particular stimulus. But this method – behaviorism – had its critics, too, and it is now seen as just one of the many ways of studying the mind and hypothesizing about how it might work.

VOICES FROM THE PAST

A man sought advice because he was troubled by voices (inaudible to anyone else) which took the form of supernatural messages. He was counseled to listen hard to the voices and pay them the attention they deserved since the gods did not waste their time with unimportant communications. Reassured by this, the man went about his business, heeding the gods, and lived a prosperous life, eventually dying in a shipwreck not far from the thriving city of Troy.

This might have been the scenario about 3,000 years ago if one interesting – if not widely accepted – theory of the nature of the human mind is to be believed. Psychologist Julian Jaynes has put forward the theory that a sense of self, a coherent consciousness or ego, only developed about 3,000 years ago. Before that he maintains that the mind was two-chambered, or bicameral, referring to the left and right hemispheres which were, he claims, not integrated. One hemisphere "spoke" and the other followed its orders, attributing the "voice" to the gods. Today a man who sought advice because he was hearing voices would probably be diagnosed as schizophrenic and be prescribed powerful drugs to treat the symptoms.

One theory of the mind was phrenology, developed by German physician Franz Gall in the 19th century. He believed the shape of the brain correlated with many emotional and temperamental qualities and that its shape could be worked out from the form of the skull. Character could thus be assessed by feeling the bumps on a person's head. Phrenological heads (**above**) were made, which mapped out where a particular quality might be felt on the head. Today the theory has no currency.

The discipline of ethology studies the behavior of animals in the natural environment, rather than in laboratory conditions. The ethological approach has been recognized by psychologists as a valuable way of understanding human behavior by comparing it with the way animals behave. It emerged in Europe and reached its height with the work of Konrad Lorenz and Niko Tinbergen in the 1970s.

Ethology tries to show how an animal's innate behavior could have evolved so as to fulfill a function and be useful for its survival. It also seeks to explain the rituals of animal courtship and aggressive behavior.

An ethological approach to sexual attraction between humans might look at it in the light of instinctive behavior patterns.

The cognitive approach originated during the 1950s and '60s when some psychologists were inspired by computers to look at the way people think or process information, that is, the sequence of mental events. Using this method, a psychologist might ask, for instance, if people perceive things in the same way that computers are programmed to. However, it is difficult to get inside the human mind to prove or disprove the cognitive psychologists' theories.

A cognitive psychologist who was looking at sexual attraction would try to analyze the mental processes taking place – taking reaction times and the subject's own analysis into account and, perhaps, creating computer simulations.

The biological approach is one of the most fruitful ways of studying the mind. Much has been learned by examining the physical substance and processes of the brain, rather than its mental processes or behavior. Neuroscientists, for instance, have revealed how certain chemicals work in the brain to affect our behavior and that different parts of the brain are linked to special tasks and abilities.

A neuroscientist studying sexual attraction might look for evidence of the effects of hormones on the brain or use scanners to monitor what is going on in particular regions when an attractive person is seen.

Behaviorism was the dominant approach to psychology from about 1920 to the mid-1950s, and it still has many advocates. It concentrates entirely on how people (and animals) behave and tries to analyze all behavior in terms of a stimulus and a response. Behaviorism was made famous by the Russian physiologist Ivan

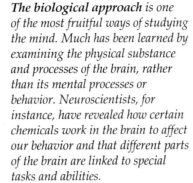

Pavlov (1849–1936), who showed how dogs could be "conditioned" to give a specific response to a particular stimulus, such as a ringing bell, which they associated with food.

B.F. Skinner (1904–90), one of its most extreme proponents, argued that all human behavior could be explained by stimulus–response relationships. In his novel Walden Two, he showed how a few powerful people could keep everyone "happy" by creating a world in which they were stimulated in the right way.

A behaviorist considering sexual attraction would look at the detail of physical response when an attractive person is seen – the way the people involved move or what happens to the eyes, for instance.

The psychoanalytic approach was pioneered by Sigmund Freud (1856–1939). The essential idea behind it is that people's psychological history explains to a large extent both the make-up of their character and their current behavior. Freud suggested that we have "conscious" thoughts of which we are aware and "unconscious" thoughts of which we are unaware, but which have a great influence on the conscious mind. He emphasized the role of childhood experiences, particularly during the first five years of life, and the importance of sexuality in forming the adult personality.

A psychoanalyst studying sexual attraction might try to analyze a person's childhood and relationship with the parents to find the origin of sexual feelings.

Gestalt psychology was a reaction in the 1920s and '30s by many Austrian and German psychologists, notably Max Wertheimer (1880–1943), against behaviorism and the way it broke everything down into units of stimulus and response. Gestalt psychologists were interested in the mind's Gestalt – its entire, integrated form – believing that the whole is greater than the sum of the parts. They were concerned with understanding mental experience and development, and concentrated on people's awareness of the world and their surroundings.

Whereas a behaviorist studying sexual attraction might try to analyze the minutiae of a person's reactions, a Gestalt psychologist would look at what the whole experience of attraction means.

See also

SURVEYING THE MIND
▶ What is a brain? 12/13

▶ Discovering the brain 14/15

▶ Comparing brains 18/19

▶ Inside the mind 20/21

▶ Electronic minds 22/23

▶ Probing the mind 26/27

INPUTS AND OUTPUTS
▶ Survival sense 122/123

FAR HORIZONS
▶ Conditioning the mind 128/129

▶ Male and female 148/149

STATES OF MIND
▶ Being aware 158/159

▶ Abnormal states 162/163

▶ Talking cures 164/165

Comparing brains

Animal studies have revealed much about how brains work and give some insights into human thinking.

Physicians and anatomists have long realized that there was a great deal to be learned about the human body by studying animals. Since they have the same or similar senses and nervous systems, and even their brains have the same basic arrangement, animals have been the subjects of countless experiments. Anatomists and, more recently, neuroscientists have used animals to learn about the human brain and nervous system. Much of our knowledge about learning and memory, for instance, comes from studies of white rats.

It was only gradually accepted, however, that animals may be able to teach us something about the human mind. Early in this century, psychologists developed the idea of behaviorism, which showed that learning in both humans and animals can be studied without worrying whether they think in the same way. It entails simply observing how a person or animal behaves in response to a particular stimulus.

Much of the research was geared to explaining how behavior is learned. But many scientists argued that some kinds of behavior were not learned but inborn, and "ethologists" studied animals' natural behavior to see if they could identify any patterns. One noticed, for instance, that newly hatched goslings will "imprint" – form an immediate attachment – to the first moving object they

see nearby. Others have discovered different forms of inborn behavior, such as "fixed-action patterns" – patterns of behavior that are repeated seemingly automatically whenever certain events occur. Although it is important to remember that such animal experiments do not teach us about human behavior directly, they provide a wealth of ideas which can guide research.

***Termites** live in a tightly regulated community with different castes that have specific roles in the nest. Humans, too, live in societies, but – unlike termites, whose role seems genetically predetermined – are, in principle, able to move between roles.*

Dogs have a good sense of smell and can track prey over long distances, since their noses are packed with scent receptors. They also communicate using smell – by scent-marking their territories with urine.

Smell receptors

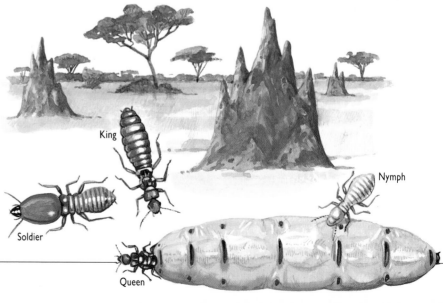

King

Nymph

Soldier

Queen

A spider has a tiny brain, but this does not mean that it cannot perform feats that humans find impossible. The delicate tracery of a spider's web is an example of animal engineering at its finest. Not only can the spider instinctively make the web – accurately positioning the supporting strands and the infill threads – but it can also tell by the vibrations made when an object hits the web whether that object is a prey insect or a useless bit of fluff.

Only in the 20th century have humans managed powered flight. But birds have been doing it for millions of years. By studying how birds use air currents, how their wings are shaped, and how they navigate, scientists and engineers have learned much about the principles of flight.

The cortex is the part of the brain where higher functions take place, so a large cortex relative to body size is seen as a sign of intelligence. If the areas of the cortices of various animals are shown by different-sized pieces of paper, a rat's cortex is the size of a postage stamp; a monkey's the size of an envelope; a chimpanzee's the size of a sheet of typing paper; and the human cortex the size of four sheets. However, the cortex of an elephant is four times as big as a human's and that of a whale is six times greater. But the human cortex is large in relation to the size of the body, while the whale's and the elephant's are both much smaller in comparison to their bulk.

Some snakes can sense heat radiation. This lets them find prey in the dark by its body heat. They use a part of the spectrum not visible to humans, which is called the infrared because it comes beyond the red end of the rainbow of colors that we are able to see.

Many bats navigate and hunt by means of a high-frequency echolocation system that uses sounds too high-pitched for us to detect. While the human ear can at best hear sound frequencies between 20 and 20,000 Hz (Hertz, or waves per second), some bats use sound frequencies of up to 200,000 Hz, when hunting insects, for instance.

See also

SURVEYING THE MIND
▶ What is a brain?
12/13

▶ The working mind
16/17

▶ Probing the mind
26/27

BUILDING THE BRAIN
▶ The gap gallery
54/55

INPUTS AND OUTPUTS
▶ Guided motion
86/87

▶ The mind's ear
98/99

FAR HORIZONS
▶ Conditioning the mind
128/129

▶ Learning to be human
144/145

STATES OF MIND
▶ The resting mind?
160/161

▶ Physical cures
166/167

Inside the mind

We think all the time, whether we are aware of it or not. But what does the process of thinking really involve?

Psychologists can easily study our behavior, the external result of our brain at work. Neuroscientists can also study the anatomy and neurochemistry of the brain, but it is far harder to see the mind in action. Experiments on animals, modern scanning techniques, and experience with computers are slowly helping to build up at least a vague picture of what happens.

Even something as simple as deciding to sit down involves a complex sequence of events. The brain receives a continual flow of information from the senses and the rest of the body – via the nerves and hormone system – which it analyzes before sending instructions out to the body. All the time, the analysis and orders are modified by feedback of more data. Thus the eyes will scan the room to look for a chair, messages go to the brain, and a seat is chosen. Motor impulses, or nerve signals, from the brain stimulate muscles to make us move across the room. Then we have to assess the angle of the chair, how far to bend, how to keep our balance. Simultaneously, we could be talking to a friend, daydreaming, or rerunning a memory of a past event. Usually all this goes on without our realizing it, yet a unique facet of the human mind is that we can be self-aware – we know we are thinking – so aspects of the process can be studied.

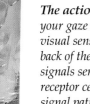

 Action Sensation Action

The action of directing your gaze to a glass gives a visual sensation of the glass at the back of the eye. This is coded into nerve signals sent to the thalamus by retinal receptor cells. The thalamus passes this nerve signal pattern to the visual cortex, where the image of the glass is perceived and recognized. It is then stored as short-term sensory memory, which holds things in the mind for a second or two even while new sensations are flooding in.

Message to brain

Perception recognition

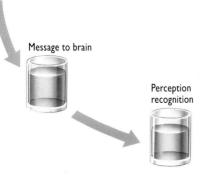

Just seeing ice cold water can make you lick your lips in anticipation, and a cascade of mental activity is set off. As you reach for the bottle,

pick it up, and pour, a stream of signals feeds back from the eye to the brain, which coordinates your movements and sends motor impulses to

the muscles to keep you moving in the right way. As you sip, many new sensations flood your brain – the taste of the water, its cool feel.

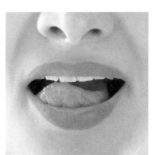

Associations spring to mind when you start thinking about what a glass of water means. The water may remind you of swimming in a clear pool, drinking cold lemonade, easing a parched tongue. Even seeing part of a bottle or its outline can trigger associations; they can be about almost anything in your experience and do not have to be directly about water. You may think, for example, about lack of water in a drought-stricken landscape.

Association comes only when the glass has been perceived and recognized. When the visual cortex "sees" the glass, the mind does a rapid check of its qualities – its shininess, its transparency, the way the water in it moves – to see if it can be identified. The idea is held in the mind for a while; like letters drawn in the air with a sparkler, the pattern of every visual image persists in the brain for a moment.

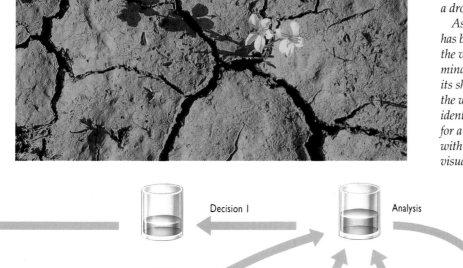

With the glass of water held in sensory memory, association takes place. Next comes analysis of the situation. Input from the hypothalamus, which monitors water balance, may follow. If the input is "thirsty" (input 1), decision 1 is made, leading to the action of drinking. The whole process then starts again and continues until at the analysis stage the input received is input 2 – "not thirsty" – and decision 2 takes place. The action decided on might be to cease drinking and direct the gaze to something else. This in turn would spark a whole new round of the sequence, which is repeated endlessly and often innumerable times a day.

Decision 1
Analysis
Input 1
Input 2
Sensory memory
Association
Decision 2

See also

SURVEYING THE MIND
▶ Discovering the brain
14/15

▶ The working mind
16/17

▶ Comparing brains
18/19

INPUTS AND OUTPUTS
▶ Guided motion
86/87

▶ A matter of taste
106/107

▶ Keeping control
116/117

▶ Dealing with drives
118/119

FAR HORIZONS
▶ The evolving mind
126/127

▶ The infinite store?
130/131

▶ Having ideas
140/141

STATES OF MIND
▶ Being aware
158/159

▶ Motivation
180/181

Electronic minds

Human brains and electronic computers seem to have some things in common. But just how alike are they?

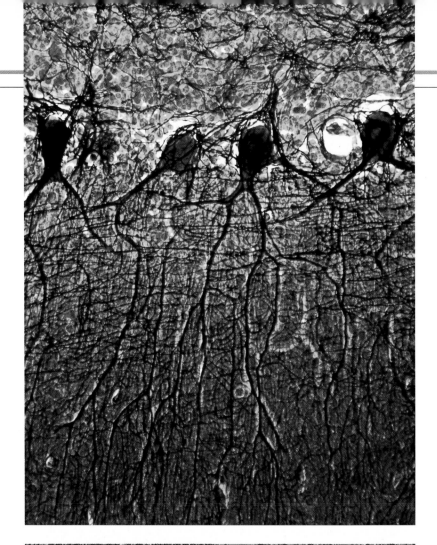

Computers that talk and have all manner of human frailties, like the chirpy R2D2 in *Star Wars* and the sinister HAL 9000 in *2001*, are, so far at least, just science fiction. But the idea that computers might one day have minds and personalities like humans is the subject of fierce scientific controversy. Equally controversial is the idea that the human brain is simply a complicated computer.

Some argue that the brain is just a biological machine – albeit a complex one. If this is so, there is no reason why we cannot, ultimately, build an equivalent electronic machine that works in exactly the same way – and because electronic impulses can travel a million times faster than nerve impulses, this computer brain could be far superior to ours. Others argue that the brain is fundamentally different from a computer. Consciousness, intuition, and things we take for granted, such as our ability to come up with ideas or use common sense, are all things that a computer will never be able to mimic, because these faculties do not work mechanically, but in a fundamentally different (and as yet unknown) way.

Nonetheless, there are certainly similarities between computers and brains, and research into the parallels has helped both those working in Artificial Intelligence (AI, the theoretical basis for a thinking computer) and neuroscientists studying the brain. But so-called bottom-up comparisons between computers and brains at the level of individual cells and transistors have actually taught us little about the brain.

The real interest in computer–brain comparison at the moment is in the top-down approach – that is, comparisons between their overall functioning. Both computers and brains involve networks of connections, so AI experts have worked with the idea of neural networks – groups of connections that mimic particular brain functions. The idea is to come up with theoretical models of networks of neurons that, for instance, learn things the same way the brain learns them. But many scientists still feel that neural networks may ultimately only help in the development of computers and not in the understanding of the brain, and especially not in the understanding of qualities of the mind such as imagination.

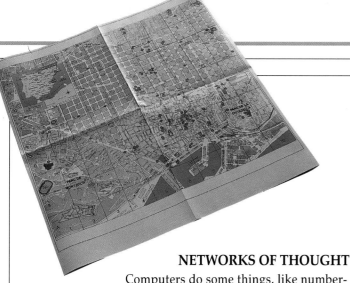

A map enables us to find out how to get from one place to another. A set of instructions using a limited set of rules can be worked out, such as first left, second right, straight ahead at the junction, and then fork left. These are not unlike the components of an algorithm, a set of rules that if applied will lead to the solution of a problem.

NETWORKS OF THOUGHT

Computers do some things, like number-crunching and playing chess – indeed anything involving rigid logic – better than humans. Yet there are things like recognizing a good picture, laughing at a joke, or cooking a meal that we find effortless at which even the most sophisticated computers fail miserably. In attempting to understand why, scientists have tried to find some logic in the brain's activity by creating neural networks.

Neural networks are designed using algorithms – sets of mathematical rules that rigidly guide the network's operation, so a particular data input will produce a particular output. A neural network is basically a way of arranging a computer's connections that will make it follow a particular algorithm. The idea is that the neural network mimics some brain function precisely, giving the same output for a given input. An important element in many networks is the ability to learn, and some can teach themselves.

One of the problems with neural networks is that it is difficult to find a single brain function that a network might mimic, since so many things in the brain work together and interact. Moreover, only rarely does the brain give an unvarying, inevitable output to a particular input in the way that a neural network must. Usually, each of the myriad inputs into the brain is mixed in with all the rest and stored for the future rather than producing a particular instant response.

Machines can be made to look similar to humans, but they cannot do many humanlike things. The mannequin above has an articulated body and can sweat and breathe, but it would have trouble playing soccer against a human.

At the basic cell level, brains initially seem similar to computers. Computers, for instance, are basically a series of tiny switches that can be set at either on or off. Neurons, or brain cells (**left, above**), also seem switchlike because they either "fire," sending an electrochemical signal, or do not. However, neurons are never simply on or off; their level of excitation and their shape dramatically alter the way they communicate with one another. Attempts to create electronic versions of neurons have not worked well. In simpler computers, which use chips (**left**), circuits are connected in series – connections are made rapidly one after the other, so a signal runs through a single circuit. In the brain, however, neurons are connected in parallel – thousands or even millions of connections are made at the same time, and the signal divides and runs through huge numbers of circuits simultaneously, giving phenomenal power.

See also

SURVEYING THE MIND
▶ What is a brain? 12/13

▶ Discovering the brain 14/15

▶ The working mind 16/17

▶ Inside the mind 20/21

BUILDING THE BRAIN
▶ To fire or not to fire? 48/49

FAR HORIZONS
▶ The infinite store? 130/131

▶ Having ideas 140/141

STATES OF MIND
▶ Being aware 158/159

Discovering through damage

Sometimes the behavior changes of a patient with brain damage give insights into the mind's workings.

Tragic as they may be for those involved, the effects of tumors, strokes, gunshots, poisonings, and other forms of brain damage have been valuable sources of information about which parts of the brain do what. Before the invention of modern imaging techniques, the region of damage could only be discovered after a patient's death. Researchers therefore had to wait for autopsy results before they could relate changes in behavior or function to the place in a person's nervous system where there was a lesion.

The process of uncovering which functions and abilities are located in specific regions of the brain gained momentum from an observation by Parisian doctor Paul Broca. In 1861 he described a patient who, because of damage to a small region of the left side of the motor cortex, now known as Broca's area, had lost the power of speech. Research continues today into all sorts of damage, including, for example, that to the parietal lobe, which causes curious gaps in the victims' knowledge or abilities. One patient with damage on the right side lost all sense of the left of his body, ignoring anything on that side. He once drew a watch with all the numbers crammed into the right side.

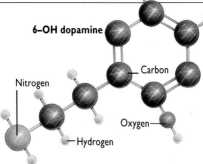

6-OH dopamine

Nitrogen — *Carbon* — *Oxygen* — *Hydrogen*

6-OH (hydroxy) DA (dopamine) *– similar to the neurotransmitter dopamine – is taken in by dopamine receptors, producing some of the symptoms of Parkinson's disease. Since 6-OH DA is taken up selectively into dopamine cells, it can be used to research their functioning.*

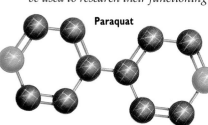

Paraquat

Paraquat *is similar in structure to MPTP. It is found in certain types of weedkiller and if ingested causes fatal lung damage. There is no known antidote for paraquat, so once enough has been absorbed it will kill. In areas where paraquat is used, there is a small statistical rise in the numbers of cases of Parkinson's disease. Its structural similarity to MPTP may account for this.*

In the 1970s *a group of people in their 20s started to develop the symptoms of Parkinson's disease. They were brought on by the impurity MPTP, in a compound made by people attempting to synthesize heroinlike drugs. MPTP selectively destroyed the dopamine cells of the substantia nigra region of the basal ganglia.*

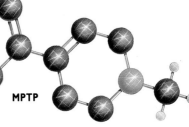

MPTP

GAGE THE DAMAGE

American railroad construction worker Phineas Gage was a member of a blasting crew. One day in 1848, while he was ramming dynamite into a hole drilled in rock, a blast occurred which sent his iron tamping rod – a yard long and 1½ inches (4 cm) wide – through his left cheek and out of the top of his head. Amazingly, he survived.

But when Gage recovered from his injuries, his behavior had altered. He suffered a dramatic change of personality: from being a man described as shrewd, well balanced, and persistent, he became moody, difficult, given to bad language, unable to plan ahead, and inconsiderate of other's feelings.

The rod had seriously damaged the frontal lobes of his cortex, thus providing good evidence that the frontal lobes control our sense of self and ability to carry out long-term plans. Surprisingly, in view of its drastic effects, about a century later deliberately destroying the frontal lobe briefly became a treatment for some intractable psychiatric illnesses.

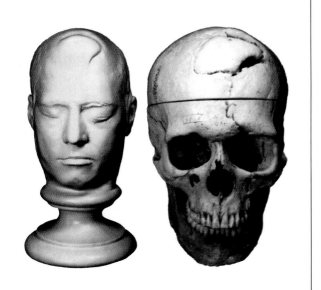

Phineas Gage's death mask *and his skull show where the metal rod that injured him entered and left his skull.*

❶ *Prefrontal lobe damage – as in the case of Phineas Gage – alters the ability of people to execute plans and can make them inconsiderate. It can also pacify them.*

❷ *Frontal lobe damage affects movement since this area contains the pre-motor and motor cortices.*

❸ *Damage to the parietal lobe of the cortex interferes with perception of touch and pain, as well as some visual functions and the knowledge of where the body is in space.*

❹ *Posterior parietal lesions can have bizarre effects, one of which is referred to as "neglect." When the left posterior parietal lobe is damaged, "right neglect" occurs. Typically, a patient treats the right side of the body and the world around it as if they did not exist. This is because each side of this region of the brain integrates inputs from the different senses to form a coherent picture of what is happening in the opposite side of the body, where the signals originated.*

❺ *Damage to any of the areas marked with diagonal lines can cause blindsight – a person can detect something visually, but cannot acknowledge that he or she can see it.*

❻ *Injury in the V5 visual (occipital) region, which is thought to play a part in the perception of motion, causes some odd symptoms. For instance, following a stroke there, a patient might see the world only as a series of static images.*

❼ *People with damage to Wernicke's region, an area in the left temporal lobe, can speak fluently but make no sense. A typical reply to a question about a vacation might run: "Oh, yes, we have done it, could be different, but nevertheless done. Go, go, gone, and however successful it still fails." The area deals with sense and comprehension of language.*

❽ *A range of symptoms is evident in patients with damage to the temporal lobe. The right temporal lobe is involved in controlling spatial tasks, and damage to a specific part of it can render someone unable to recognize faces, even those of close family members. A farmer with damage to much the same area could recognize his family, but lost the ability to recognize his sheep. Damage to other parts of the temporal lobes can result in dramatic hallucinations or a loss of memory for any subsequent events.*

❾ *Patients with lesions in Broca's area can sometimes speak, but the speech is labored and halting. Ask them about a holiday and they might reply "Ho, ho, holiday, like…eat turkey…people…good." This region controls grammar and vocalization.*

See also

SURVEYING THE MIND
▶ Discovering the brain
14/15

▶ Probing the mind
26/27

▶ Brain maps
30/37

BUILDING THE BRAIN
▶ Recovering from damage
64/65

▶ The aging brain
76/77

INPUTS AND OUTPUTS
▶ Levels of seeing
92/93

▶ Active vision
94/95

FAR HORIZONS
▶ The infinite store?
130/131

▶ Parallel minds
138/139

STATES OF MIND
▶ Emotional states
154/155

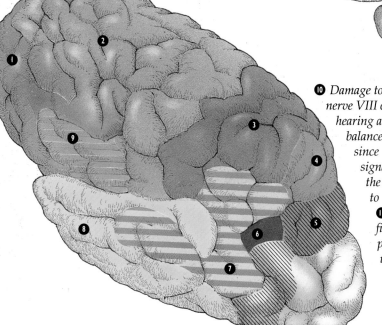

❿ *Damage to cranial nerve VIII causes hearing and balance problems since it carries signals from the inner ear to the brain.*

⓫ *The cerebellum fine-tunes motion, so problems here can cause unsteadiness and coarse movements.*

Probing the mind

Investigating the physical properties of the brain reveals much about how it works.

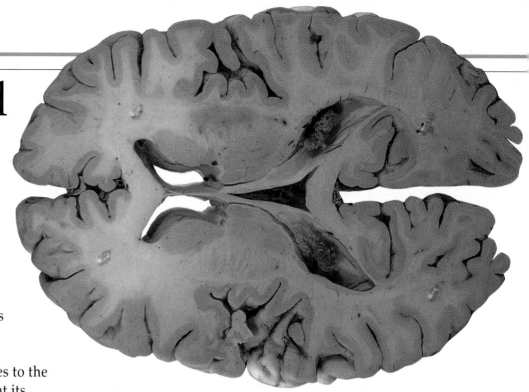

Only comparatively recently has the true scale of the complexity of the human brain emerged. It has around 100 billion nerve cells, or neurons, each connected to hundreds, sometimes thousands, of other cells. Unraveling how this vast network operates has been one of the great challenges of the 20th century. And today, scientists have a number of techniques at their disposal.

Until individual neurons were first stained and directly observed using powerful microscopes, clues to the brain's workings could be gained only by looking at its anatomical structure. But scientists can now use techniques, such as the electroencephalogram (EEG) machine, first developed in the 1930s, that measure the minute fluctuating wavelike electrical signals that result from circuits of neurons firing in unison.

Using laboratory animals, researchers can remove parts of the brain or damage them to see the effect on the behavior or function of the living animal. Or by inserting tiny recording electrodes deep into living tissue, they can home in on individual neurons and record their electrical activity. By keeping thin slices of brain alive outside of the body, scientists can use electrical probes to examine the brain's workings. They can control the environment of isolated populations of neurons and, for example, measure their response to drugs that they would not normally come into contact with. The responses are frequently registered using an electrode placed inside the neuron.

If you had never seen a basketball game before, *looking at just one player by himself would make it difficult to figure out what he was doing, what the rules are, and what his role was.*

Similarly, techniques that study the activity of just one brain cell can tell something about the way an individual cell reacts to a stimulus, but give no idea of the

big picture. For instance, only truly vast numbers of brain cells working together as a unit can process our visual inputs of the outside world on the way to perception, or make our muscles work. And some of the more complicated brain functions, such as making decisions, involve precisely coordinated networks of cells from half a dozen different brain areas.

MAKING WAVES

Using electroencephalogram (EEG) machines, the tiny voltage changes (around 10 microvolts) that result from the synchronized firing of large groups of neurons can be picked up by electrodes placed on the scalp. Because the current involved is so small, it is amplified before being traced out on a sheet of paper where the voltage changes appear as a series of waves. There are a number of wave patterns – actually a record of the difference in potential between two points on the scalp – which are named after letters of the Greek alphabet.

So alpha, with a frequency of 9 to 11 waves per second, appears when our eyes are shut and we are relaxed. Beta is faster and indicates alertness; delta and theta are slower and linked with stages of sleep. EEG readings can show what stage of sleep a person is in and, some claim, if the subject is lying. They can also distinguish types of epilepsy.

EEG waves produce their typical wave pattern because of the way that groups of neurons alternately excite and inhibit one another. If one group of neurons is excited by a stimulus, they fire when it appears, triggering a nearby group whose job it is to reduce activity; as things quiet down, the inhibitory neurons turn themselves off, allowing the excitatory ones to fire up again. The size of an EEG wave reveals how many cells are synchronizing their impulses; a larger wave indicates more synchronization.

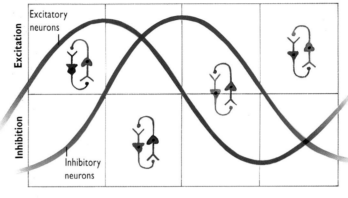

Recordings from individual neurons form the basis of our understanding of the brain. Typically these recordings – obtained by placing a small electrode outside the neuron – detect the tiny pulses of electricity that it produces when it "fires," sending messages to other neurons. Much has been learned, for instance, about the different sensory systems in the brain – especially the visual system – by monitoring what individual neurons do in response to different external sensory stimuli.

Surgical operations have provided many insights into brain functions. If the corpus callosum, the net of 100 million nerve fibers linking the two cerebral hemispheres, is cut to relieve epilepsy, for example, two half-brains are created, each governing the opposite side of the body. By studying these "split-brain" patients, doctors have been able to determine the differing roles of the two sides of the brain.

If the hippocampus is destroyed (when a tumor is being removed, for example), the ability to lay down new memories is lost, although people can remember details of their past life and how to speak. But they do not recognize people they have met only hours earlier and can read a page repeatedly without knowing they have seen it before.

Taking a slice of brain and then experimenting on it is a favorite technique for researchers. The slice from the brain of an animal such as a rat can be used in several ways. It can be stained so that individual neurons and their connections show up. Or it can be kept in a special solution that sustains the cells for a few hours; this allows researchers to experiment on living brain tissue. In autoradiography, a radioactively tagged substance is injected into the animal before the slice is taken. An image (**left**) obtained using a film sensitive to the radiation shows how that substance is used by regions of the brain.

Scientists can not only detect signals of the electrical pulses of individual brain cells, but they can also see how electricity is involved in minute parts of the cells themselves. In a technique known as patch clamping, microfine glass pipettes are used to record the activity of a single ion channel in the wall of a brain cell, responsible for the flow of electricity. The tip of the pipette is placed against the cell membrane and a tight seal created. The movement of the ions through the channel can be measured in terms of the current that this creates.

See also

SURVEYING THE MIND
▶ Comparing brains
18/19

▶ Discovering through damage
24/25

▶ The view from outside
28/29

BUILDING THE BRAIN
▶ The brain's cells
44/45

▶ The electric cell
46/47

▶ To fire or not to fire?
48/49

▶ On or off
56/57

▶ Recovering from damage
64/65

INPUTS AND OUTPUTS
▶ On the scent
108/109

▶ Rhythms of the mind
120/121

FAR HORIZONS
▶ Active memory
132/133

The view from outside

Modern imaging techniques cannot tell what you are thinking, but they can tell where it happens in the brain.

A human brain is a delicate thing, and the challenge to researchers studying it is to discover what it is doing without causing it any damage. The first pictures of the living brain were taken with X-rays in 1917, but they gave only general information about structure since X-rays are not able to show much detail in soft tissues such as those that make up the nervous system. Only about 50 years later, with the arrival of the computerized axial tomography (CAT) scanner, could X-rays be used to produce the first detailed images of the brain.

But CAT images cannot show the brain in action. Today, however, there are a number of alternative techniques that make this possible. These include positron emission tomography (PET), functional magnetic resonance imaging (fMRI), and magnetoencephalography (MEG). None of these uses potentially harmful X-rays. Currently, PET scans can give information on what is happening at 30-second intervals over

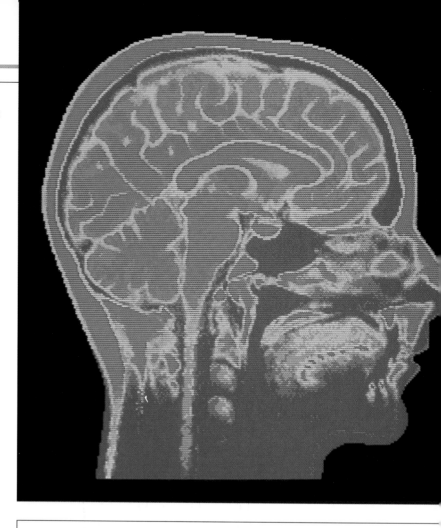

Positron emission tomography (PET) involves injecting into the blood a tiny amount of a radioactive substance, attached to molecules that are absorbed during brain activity.

This gives off gamma rays that can be recorded by sensors and analyzed by computers to build up a picture of where in the brain increased use of the molecules is taking place.

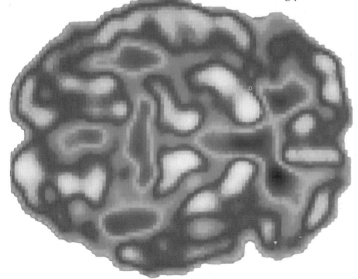

ROENTGEN'S REVEALING RAYS

Discovered by German physicist Wilhelm Roentgen in 1895, X-rays made it possible to obtain images of structures inside the body without surgery. Although it was a breakthrough, X-rays do have limitations. They can only show large-scale features such as tumors, since the minutiae of soft tissues like the brain do not show up well, and they cannot indicate depth or the different layers within an organ. They also cannot always differentiate between healthy and diseased tissues.

Another limitation of ordinary X-rays is that they can reveal only what is in the path of a single burst of radiation. The same is not true of CAT scans, which provide detailed images of the brain's soft tissues. And by putting a number of CAT scan "slices" together, a 3-D image can be assembled. However, although CAT scanning is a great advance, like all X-ray techniques it can only take snapshots of structure and does not show changes in the brain's activities.

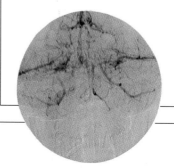

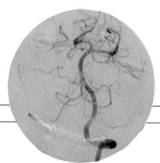

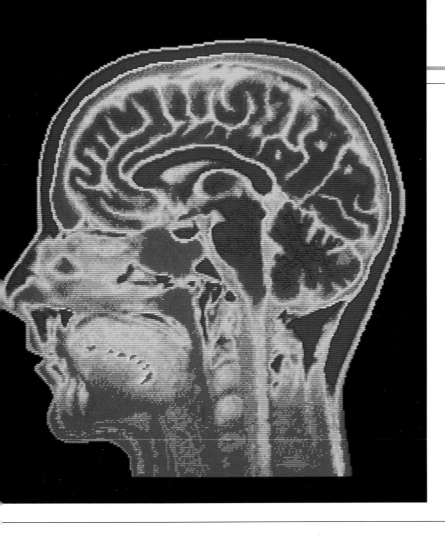

Magnetic resonance imaging (MRI) works because atoms in body molecules can be made to behave like tiny magnets if they are put in a strong magnetic field. A beam of radio waves fired at them will make them resonate and give off radio signals of their own. Sensors detect these signals, and the information can be made into an image by a computer.

an area of about $\frac{1}{13}$ inch (2 mm). But this is still not enough to detect brain changes that may last only a fraction of a second. While fMRI is somewhat better than PET, it still cannot show the extremely fast-moving events that take place in the brain. However, the new MEG technique makes this possible to some extent. Active groups of nerve cells generate tiny electric currents, which can be shown in electroencephalogram (EEG) recordings; MEG works by detecting the magnetic fields these tiny currents generate. The computer-processed images produced make it possible to record the site of brain activity by the millisecond.

See also

SURVEYING THE MIND
Discovering the brain
14/15

Probing the mind
26/27

Brain maps
30/37

BUILDING THE BRAIN
Support and protection
42/43

Recovering from damage
64/65

Food for thought
66/67

FAR HORIZONS
Word power
136/137

STATES OF MIND
The failing mind
176/177

Computerized axial tomography (CAT) scans combine X-rays and computer technology to show details of the brain itself. Information from a number of X-ray sources placed around the head is digitized and then reconstructed to give a cross-sectional view of the brain in any plane that is wanted (right).

The blood supply to and within the brain can be seen using cerebral arteriography (bottom left and center). A radio-opaque dye – which does not allow X-rays to pass through it and so shows up on an X-ray image – is injected into a main artery supplying the brain. As it diffuses into the blood, X-rays are taken. They reveal the distribution of the dye, which shows the size and shape of the vessels. The X-rays can be used to find the site of any narrowing, blockage, or displacement in blood vessels. By contrast, a simple X-ray of the head (bottom right) shows little about the brain structure, since bone absorbs most of the X-rays.

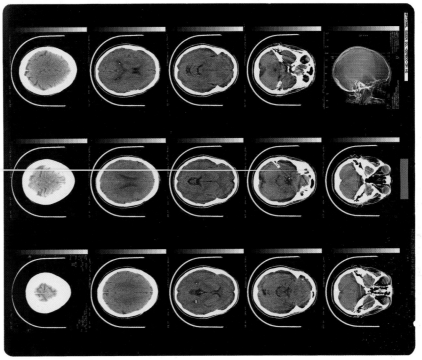

Brain map 1

There is far more to the convolutions of the brain than meets the eye.

We all know where the brain is and have a rough idea of what it looks like. But once we start exploring the brain closely, everyday knowledge soon proves inadequate. This is the case especially when the outer layers of the brain are stripped away to reveal the maze of internal structures.

Anatomists have identified numerous individual regions that are distinct from one another in any number of ways, in terms of both appearance and function. They may, for example, vary in color, have a different texture, or be encased in a self-limiting membrane, or they may be responsible for one highly specific brain function.

The 12 pairs of cranial nerves, which branch directly from the brain itself, can be seen when the brain is viewed from below. Several of these nerves take signals from the sense organs to the brain, but others connect the brain with body organs such as the heart and lungs.

When the brain is viewed from above, the wrinkled mass of the cerebral hemispheres is revealed. The hemispheres are slightly narrower at the front than at the back.

The brain is dominated by the cerebrum, with the cerebellum (little brain) below. It is connected to the body via the spinal cord, together with its nerves, and via the cranial nerves.

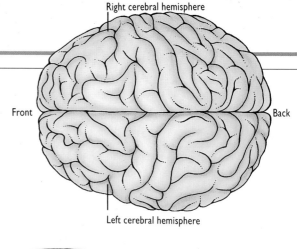

Right cerebral hemisphere

Front

Back

Left cerebral hemisphere

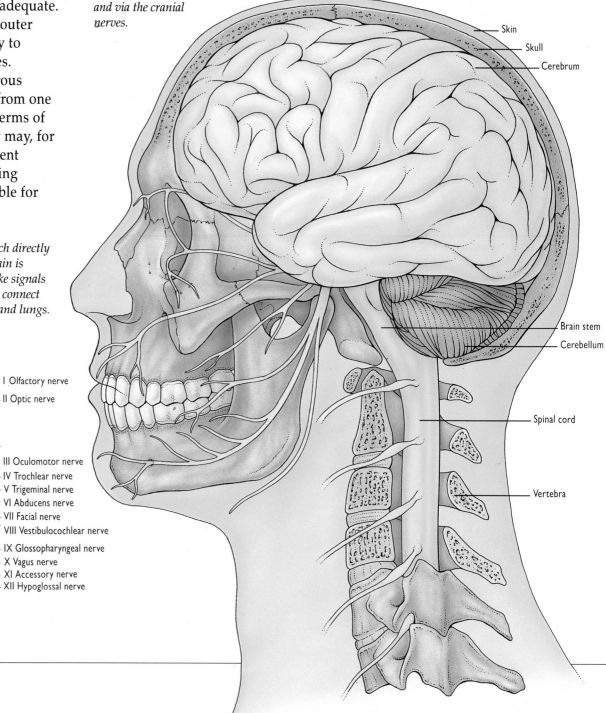

Skin

Skull

Cerebrum

Brain stem

Cerebellum

Spinal cord

Vertebra

I Olfactory nerve
II Optic nerve
III Oculomotor nerve
IV Trochlear nerve
V Trigeminal nerve
VI Abducens nerve
VII Facial nerve
VIII Vestibulocochlear nerve
IX Glossopharyngeal nerve
X Vagus nerve
XI Accessory nerve
XII Hypoglossal nerve

In the map on the right, the central region shows a cross section more or less through the center line of the brain, from front to back, as far out as the corpus callosum. The outer region represents a slightly off-center cross section, cut in the same direction.

The relatively thin outer layer of each cerebral hemisphere is the cerebral cortex, which is packed with the cell bodies of nerve cells and their many local interconnections. Beneath that the cerebrum consists of the message-carrying fibers, or axons, of nerve cells. Regions of the cortex are linked to one another by short association fibers, which take signals to and from adjacent areas, and by long association tracts, which take signals from lobe to lobe of the cortex. Projection fibers take signals between the left and right hemispheres via the corpus callosum, and from the outer cortex down to inner brain regions such as the thalamus and hypothalamus, and on to the body.

Projection fibers
Short association fibers
Choroid plexus
Pineal gland
Superior colliculus
Inferior colliculus
Long association tract
Cerebral cortex

Corpus callosum
Fornix
Septum pellucidum
Thalamus
Midbrain
Cerebellum
Optic chiasma
Pituitary
Anterior commissure
Hypothalamus
Interthalamic connection
Pons
Medulla

Choroid plexus
Corpus callosum

Internal capsule
Third ventricle
Substantia nigra
Mammillary body
Fourth ventricle

Pons

Caudate nucleus
Lateral ventricle
Fornix
Thalamus
Putamen
Globus pallidus
Optic tract
Medial longitudinal fasciculus

Olivary body

Pyramidal decussation

The map on the left shows a cross section of the brain taken from left to right approximately from ear to ear. It reveals many of the inner brain structures which are normally hidden, including the thalamus, caudate nucleus, putamen, and globus pallidus. Also shown are some of the fluid-filled cavities known as ventricles.

The nerve fibers connecting the two cerebral hemispheres are in the corpus callosum, and the fibers that link the cortices to other brain regions and the body are in the internal capsule and the medial longitudinal fasciculus. Most nerve fibers from the brain cross over from one side to the other at the pyramidal decussation at the top of the brain stem.

See also

SURVEYING THE MIND
▶ Discovering through damage 24/25

▶ The view from outside 28/29

▶ Brain maps 32/37

BUILDING THE BRAIN
▶ Support and protection 42/43

▶ The brain's cells 44/45

▶ The brain plan 68/69

INPUTS AND OUTPUTS
▶ Body links 80/81

▶ Guided motion 86/87

▶ The long junction 82/83

▶ The big picture 114/115

▶ Dealing with drives 118/119

FAR HORIZONS
▶ Parallel minds 138/139

Brain map 2

Explore the details of the brain's core and cortex.

The naming of parts of the brain can at first seem confusing, particularly since much of the terminology is either in Latin or Greek (or derived from them). This is because these languages provide universal terms which can be used by scientists regardless of their mother tongue. However, once you have mastered the meaning of the most common words, it becomes much easier to understand the terms.

Ten important Latin words are used as directional terms: superior means above or toward the top; inferior means below or toward the bottom; ventral means underneath or lower; dorsal means upper or higher; anterior and rostral both mean toward the front; posterior and caudate mean toward the back; medial means toward the middle; and lateral away from the middle. Three other words are also often used. Septum describes a wall between two cavities; a gyrus is a ridge on the cerebral cortex; and a sulcus is a groove – on the cerebral cortex, sulci are the grooves between gyri. Many words describe shape, hence the hippocampus (from the Greek words for horse and sea monster), which resembles a sea horse in cross section.

With the overlying cerebral hemispheres removed, *the regions in the interior of the brain become visible. For clarity (**right**), the two structures labeled lentiform nucleus have been pulled away from the thalamus, which they surround. The eyes are shown since they are effectively outstations of the brain: they house the retinas, which contain nerve cells, or neurons. They send partly processed signals back to the brain via the optic nerves, chiasma, and tract, which arrive at the lateral geniculate bodies, or nuclei, of the thalamus.*

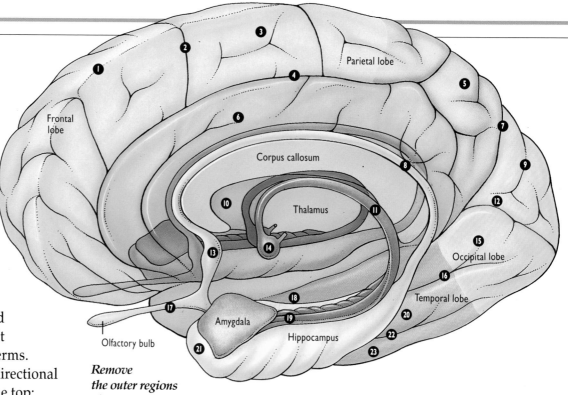

Remove *the outer regions of one cerebral hemisphere, and normally hidden inner structures are revealed. They include some of the components of the limbic system, which is involved in emotion: the parahippocampal gyrus (**18**), the cingulate gyrus (**6**), the hippocampus, the amygdala, the fornix (**11**), the mammillary body (**14**), and parts of the thalamus, septum (**10**), and lower frontal lobe.*

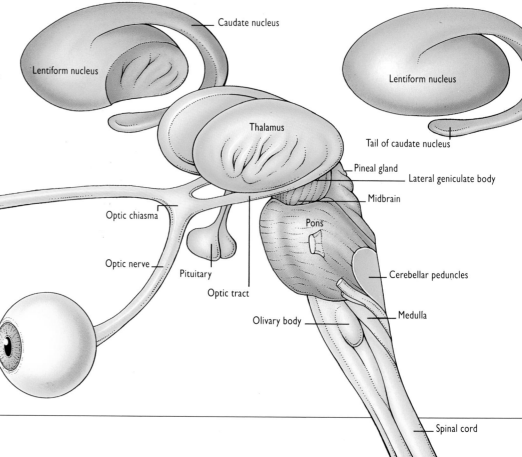

1. Medial frontal gyrus
2. Central sulcus
3. Paracentral lobule
4. Cingulate sulcus
5. Precuneus
6. Cingulate gyrus
7. Parietal occipital sulcus
8. Dorsal fornix
9. Cuneus
10. Septum pellucidum
11. Body of fornix
12. Calcarine sulcus
13. Paraterminal gyrus
14. Mammillary body
15. Lingual gyrus
16. Collateral sulcus
17. Olfactory tract
18. Parahippocampal gyrus
19. Dentate gyrus
20. Medial occipital temporal gyrus
21. Uncus
22. Occipital temporal sulcus
23. Lateral occipital temporal gyrus
24. Superior frontal gyrus
25. Superior frontal sulcus
26. Medial frontal gyrus
27. Precentral sulcus
28. Precentral gyrus
29. Central sulcus
30. Postcentral gyrus
31. Postcentral sulcus
32. Angular gyrus
33. Inferior frontal gyrus
34. Inferior frontal sulcus
35. Lateral cerebral sulcus
36. Transoccipital sulcus
37. Lateral occipital sulcus
38. Medial temporal sulcus
39. Superior temporal gyrus
40. Superior temporal sulcus
41. Medial temporal gyrus
42. Medial temporal sulcus
43. Inferior temporal gyrus

The sulci and gyri of each person's brain are often in unique positions, so any label is therefore approximate.

See also

SURVEYING THE MIND
▶ Discovering through damage 24/25

▶ The view from outside 28/29

▶ Brain maps 30/37

BUILDING THE BRAIN
▶ Support and protection 42/43

▶ The brain's cells 44/45

▶ The brain plan 68/69

INPUTS AND OUTPUTS
▶ Body links 80/81

▶ Guided motion 86/87

▶ The long junction 82/83

▶ The big picture 114/115

▶ Dealing with drives 118/119

FAR HORIZONS
▶ Parallel minds 138/139

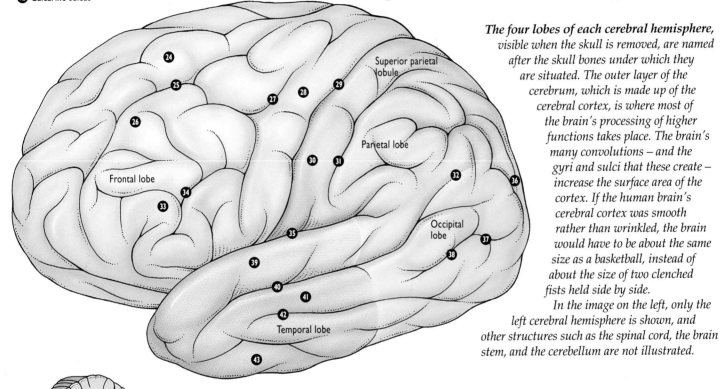

Superior parietal lobule

Parietal lobe

Frontal lobe

Occipital lobe

Temporal lobe

Cerebellar peduncles

Cerebellum

The four lobes of each cerebral hemisphere, visible when the skull is removed, are named after the skull bones under which they are situated. The outer layer of the cerebrum, which is made up of the cerebral cortex, is where most of the brain's processing of higher functions takes place. The brain's many convolutions – and the gyri and sulci that these create – increase the surface area of the cortex. If the human brain's cerebral cortex was smooth rather than wrinkled, the brain would have to be about the same size as a basketball, instead of about the size of two clenched fists held side by side.

In the image on the left, only the left cerebral hemisphere is shown, and other structures such as the spinal cord, the brain stem, and the cerebellum are not illustrated.

The cerebellum – which means, literally, "little brain" – is a distinct structure found tucked away at the rear base of the whole brain. It is attached to the brain stem by the stalklike cerebellar peduncles. Like the cerebrum, the cerebellum has a lobed structure with an outer layer of gray matter made up of nerve cell bodies and their interconnections. This outer layer is called the cortex, and it includes nuclei (bunches of neuron cell bodies and synapses) and nerve tracts (groups of fibers or axons). The basic structure of the cerebellum is evolutionarily conservative – that is, it has changed little through millions of years of evolution. The cerebella of different vertebrates – creatures with backbones – thus bear a strong resemblance to one another, unlike the cerebra, which can look extremely different.

Map of functions

Each part of the brain contributes to the overall working of the whole.

Some brain functions are neatly sited in a clearly defined region, as is the case with the hypothalamus, which is known to control certain autonomic (or automatic) body processes. But other brain functions seem to be spread across many areas, which is especially true of the cerebral cortices.

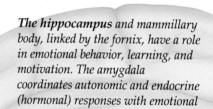

Parietal lobe

Frontal lobe

Occipital lobe

Temporal lobe

The corpus callosum (below and below right) is a bundle of nerve fibers linking the cerebral hemispheres. The caudate nucleus along with the putamen and the globus pallidus, which together make up the lentiform nucleus, have a role in the control of movement.

The hippocampus and mammillary body, linked by the fornix, have a role in emotional behavior, learning, and motivation. The amygdala coordinates autonomic and endocrine (hormonal) responses with emotional states and is involved with emotions generally. The septum pellucidum merges into the septal nucleus, which is thought to govern the level of emotional responses. Autonomic body functions, such as temperature regulation and water balance, are handled by the hypothalamus, which also links with the pituitary and, through this gland, controls the body's endocrine system.

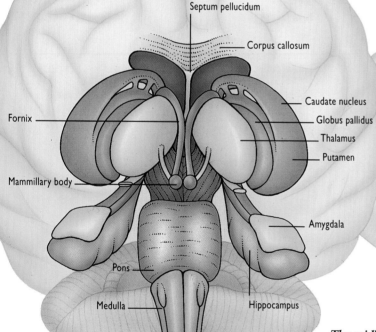

Septum pellucidum

Corpus callosum

Caudate nucleus

Globus pallidus

Thalamus

Putamen

Fornix

Mammillary body

Amygdala

Pons

Medulla

Hippocampus

Spinal cord

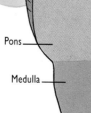

Corpus callosum

Septum pellucidum

Thalamus

Hypothalamus

Pituitary

Midbrain

The midbrain helps control sensory and motor functions such as eye movement and coordination of the visual and auditory reflexes. The pons links the medulla and the midbrain, conveys information from the cerebrum to the cerebellum, and has some respiratory control. The medulla has centers that control heart and breathing rates and blood pressure, as well as coughing, sneezing, swallowing, and vomiting.

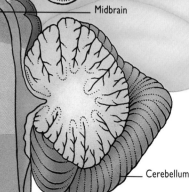

Pons

Medulla

Cerebellum

Spinal cord

The cerebral cortex can be divided by function into various regions, as shown in the colored and numbered areas on these views of the brain (**left** and **below**).

The blue dotted lines trace Brodmann's areas, which are based on differences in nerve cell arrangement and structure in the cortex.

1 This region is mainly involved with motor functions. Toward the rear is the primary motor cortex where precise muscle-moving signals originate. Toward the front is the premotor cortex, which is devoted to initiating and sequencing movements.

2 The sensations of touch, pain, and temperature are dealt with here. The receiving area for incoming signals is sited toward the front; at the rear, in the somatosensory association cortex, these somatosensory perceptions are integrated to produce an understanding of the location of the body in space.

3 Some of the higher intellectual functions, along with planning and intention, are thought to be found here. The wish to move might originate in this area, which then activates the motor regions directly behind it to carry out the detailed work of coordinating all the many muscles in the right order to initiate action.

4 Association areas are some of the many areas in the brain that have no specific role or clear function.

5 This scattered collection of regions plays a part in emotion, mood, and general behavior. The areas at the front of the frontal lobe are instrumental in the higher intellectual functions, while the region at the front of the temporal lobe is probably involved in imaging and in complex memories. Areas at the base of the frontal lobe are thought to inhibit emotions and those above the corpus callosum could add appropriate emotional responses to sensory experiences.

6 This area not only deals with olfactory functions, but also has a role in emotion and mood. The part in the temporal lobe is important in memory and olfaction, or detecting smells.

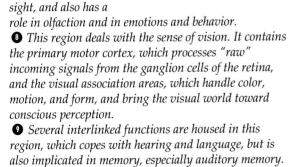

Frontal lobe

Parietal lobe

Occipital lobe

Temporal lobe

7 This cortical region deals with sensory functions of touch, taste, and sight, and also has a role in olfaction and in emotions and behavior.

8 This region deals with the sense of vision. It contains the primary motor cortex, which processes "raw" incoming signals from the ganglion cells of the retina, and the visual association areas, which handle color, motion, and form, and bring the visual world toward conscious perception.

9 Several interlinked functions are housed in this region, which copes with hearing and language, but is also implicated in memory, especially auditory memory.

The cerebellum (**right**) has a role in movement, balance, and eye movements handled by the flocculonodular lobe. The vermis and intermediate part of the hemisphere deal with the execution of movements, while the lateral part is instrumental in planning them.

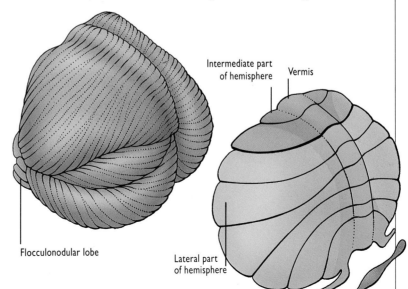

Intermediate part of hemisphere

Vermis

Flocculonodular lobe

Lateral part of hemisphere

Flocculonodular lobe

See also

SURVEYING THE MIND
▶ Discovering through damage 24/25

▶ The view from outside 28/29

▶ Brain maps 30/37

BUILDING THE BRAIN
▶ Support and protection 42/43

▶ The brain's cells 44/45

▶ The brain plan 68/69

INPUTS AND OUTPUTS
▶ Body links 80/81

▶ Guided motion 86/87

▶ The long junction 82/83

▶ The big picture 114/115

▶ Dealing with drives 118/119

FAR HORIZONS
▶ Parallel minds 138/139

Map of chemicals

A range of chemicals – neurotransmitters – play key roles in the functioning of the brain when it comes to passing nerve signals.

Neurotransmitters are chemicals that take a nerve signal across the so-called synaptic gap between a sending nerve cell, or neuron, and a receiving one. On the receiving neuron are receptors into which the neurotransmitters fit like a key in a lock. Once a neurotransmitter is bound to its specific receptor, the likelihood of the receiving cell "firing" to send its own message is affected. Some neurotransmitter–receptor systems make receiving cells more likely to fire (they are excitatory), whereas others make them less likely to fire (they are inhibitory).

A sending nerve cell releases neurotransmitters from the end of its axon, a signal-carrying fiber. The axon can link with receiving cells locally or extend to more remote receiving cells. So a nerve cell might have its cell body in the brain stem, but fibers that reach cells in the cerebral cortex. The location in the brain of the nerve cell bodies and projection fibers of six important neurotransmitters is shown here.

❶ Caudate nucleus
❷ Putamen
❸ Fornix
❹ Stria terminalis
❺ Stria medullaris
❻ Nucleus interstitialis striae terminalis
❼ Habenula
❽ Subthalamic nucleus
❾ Septum
❿ Mammillothalamic tract
⓫ Globus pallidus
⓬ Superior colliculus
⓭ Midbrain gray matter
⓮ Anterior commissure
⓯ Preoptic nucleus
⓰ Dorsal diagonal nucleus
⓱ Paraventricular nucleus
⓲ Habenulo-interpeduncular tract
⓳ Dorsal longitudinal fasciculus
⓴ Ventral tegmental area
㉑ Hypothalamus
㉒ Raphé nuclei
㉓ Cell group A13
㉔ Cell group A11
㉕ Substantia nigra
㉖ Cell group A14

㉗ Meynert's nucleus
㉘ Ventral diagonal nucleus
㉙ Interpeduncular nucleus
㉚ Anterior olfactory nucleus
㉛ Olfactory tubercle
㉜ Infundibular nucleus
㉝ Retrorubral area
㉞ Locus coeruleus
㉟ Dorsal tegmental nucleus
㊱ Lateral parabrachial nucleus
㊲ Superior central nucleus
㊳ Suprachiasmatic nucleus
㊴ Supraoptic nucleus
㊵ Amygdala
㊶ Hippocampus
㊷ Pons
㊸ Lateral lemniscal nuclei
㊹ Vestibular nucleus
㊺ Trigeminal nerve nucleus
㊻ Cochlear nerve
㊼ Nucleus solitarius
㊽ Dorsal vagal nerve nucleus
㊾ Parabrachial nuclei
㊿ Spinal nerve
51 Olivary nucleus

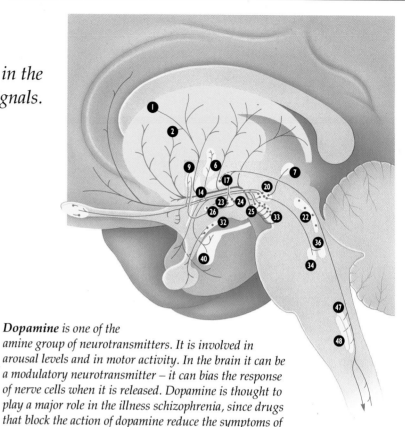

Dopamine *is one of the amine group of neurotransmitters. It is involved in arousal levels and in motor activity. In the brain it can be a modulatory neurotransmitter – it can bias the response of nerve cells when it is released. Dopamine is thought to play a major role in the illness schizophrenia, since drugs that block the action of dopamine reduce the symptoms of the illness, and drugs that enhance its action can induce symptoms similar to those of schizophrenia.*

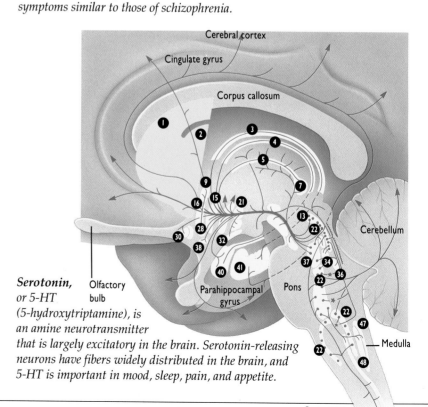

Serotonin, *or 5-HT (5-hydroxytryptamine), is an amine neurotransmitter that is largely excitatory in the brain. Serotonin-releasing neurons have fibers widely distributed in the brain, and 5-HT is important in mood, sleep, pain, and appetite.*

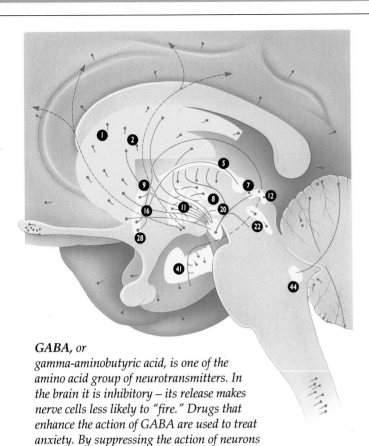

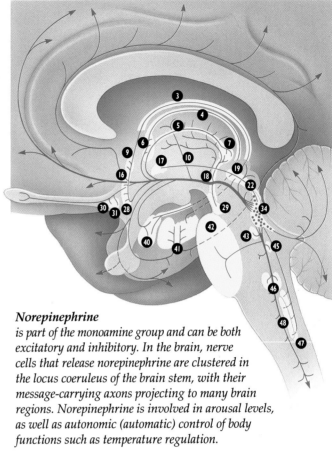

GABA, or
gamma-aminobutyric acid, is one of the amino acid group of neurotransmitters. In the brain it is inhibitory – its release makes nerve cells less likely to "fire." Drugs that enhance the action of GABA are used to treat anxiety. By suppressing the action of neurons involved specifically in the emotions, these drugs bring anxiety levels down.

Norepinephrine
is part of the monoamine group and can be both excitatory and inhibitory. In the brain, nerve cells that release norepinephrine are clustered in the locus coeruleus of the brain stem, with their message-carrying axons projecting to many brain regions. Norepinephrine is involved in arousal levels, as well as autonomic (automatic) control of body functions such as temperature regulation.

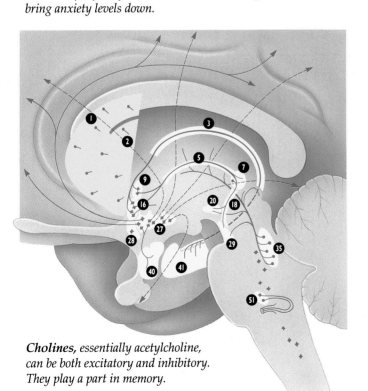

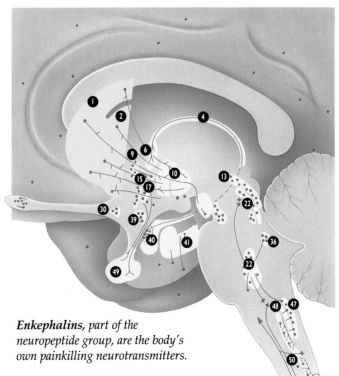

Cholines, essentially acetylcholine, can be both excitatory and inhibitory. They play a part in memory.

Enkephalins, part of the neuropeptide group, are the body's own painkilling neurotransmitters.

See also

**SURVEYING
THE MIND**
▶ Probing
the mind
26/27

▶ Brain maps
30/35

**BUILDING
THE BRAIN**
▶ Crossing
the gap
52/53

▶ On or off
56/57

▶ Unlocking
the gate
58/59

▶ Discovering
transmitters
60/61

▶ Remaking
the mind
62/63

**INPUTS AND
OUTPUTS**
▶ Feeling pain
104/105

▶ Rhythms of
the mind
120/121

**STATES OF
MIND**
▶ Physical cures
166/167

▶ Feeling low
168/169

▶ The mind adrift
170/171

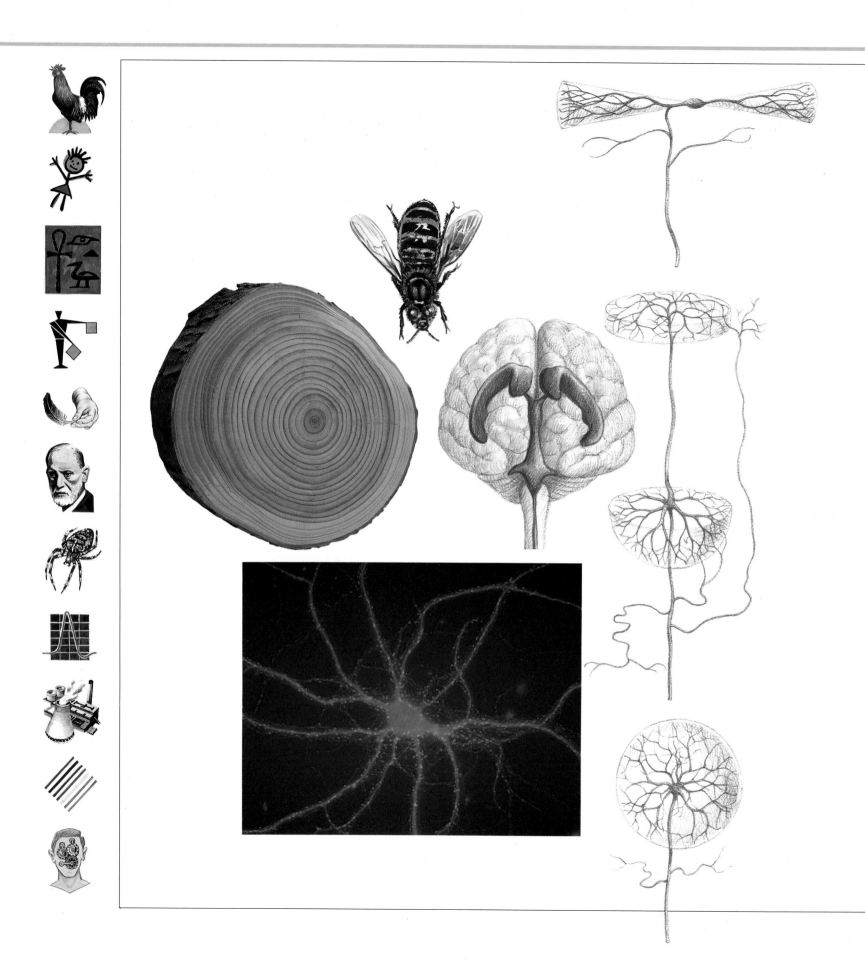

Building the Brain

O*n delving into the workings of the brain, it is possible to be overwhelmed by its layers of complexity. There are, for instance, as many as 100 billion individual neurons, each one connected to thousands of others. They constantly communicate with each other, sending up to 300 signals a second, which undergo transformations from electrical to chemical and back to electrical again. A neuron can receive messages simultaneously from thousands of other neurons; it can be bombarded with dozens of different messenger chemicals, each of which may have a different effect. All the time, the brain is changing as it adapts to new inputs, learns new skills, and lays down memories. Certain key principles can be applied across this bewildering mass of information. They apply in particular to how brain cells communicate with one another, and they are the key to understanding the brain.*

*Left (**clockwise from top**): neuron shapes; close-up on the neuron; cells like tree rings; neurons link like a bee finds its way; brain bath. **This page (top)**: learned links; (**left**) the brain's beginnings.*

Building blocks

The brain contains nerve cells, highly specialized members of the body's vast cellular community.

Life is based on the cell. This microscopic entity, averaging 20–50 microns (tens of thousandths of an inch) across, is the smallest unit that can be defined as "alive." An organism such as an amoeba has all the features of life packed into just one cell, enabling it to feed, move, grow, and reproduce. A roundworm the size of a small piece of cotton thread is composed of a few thousand cells, but unlike the amoeba's single cell, these cells cannot survive alone. They are specialized – they differ in structure and function, with each type carrying out a particular task. They depend on each other to keep the whole organism, the worm, alive.

The human body is far more complex still – it is made up of more than 50 trillion cells, of at least 200 main kinds. The different body cells work together as a vast living cooperative. To delve into the human mind, it is necessary to focus on one type of cell in particular – the nerve cell, or neuron. This is the characteristic cell of nervous tissue, including the brain – and so it could be regarded as the building block of the mind.

While neurons have most of the basic parts that other cells have, their specializations lie in their spidery shape – which allows them to connect and thus communicate with one another – and their excitable nature. Their excitability means that under certain conditions neurons can generate and

The brain is built from cells, but having more cells – a bigger brain – does not, on average, make someone smarter. If you could compare the brains of yourself, a great thinker such as Albert Einstein (left), and one of your less bright friends, you would probably see no difference in size or outward appearance. Average brain weights are 3 pounds (1.35 kg) for men and 2¾ pounds (1.25 kg) for women. Since men are generally larger than women, the brain to body ratio weight is slightly greater in females.

convey tiny pulses of electricity along their membranes. These pulses are nerve signals, also referred to as action potentials. If neurons are the cellular building blocks of the brain and nervous system, nerve signals are the data – the information carried around and stored in the system.

Little of this would be understood were it not for generations of research in the areas of anatomy, microscopy, physiology, and medicine. A major breakthrough came in 1873, when Italian physician and medical researcher Camillo Golgi (1844–1926) discovered that stains, or dyes, containing silver showed nerve cells clearly under the microscope. Previously, no one had been able to follow the twists and turns of their spidery projections as they snaked between other cells and tissues – they were too fine and impossible to see, even at high magnification.

The silver stains showed the neurons as dark lines and began to reveal the previously secret details of the nerve net. Neurology, the study of the brain and nerves, took a leap forward. In recognition of his efforts, Golgi was awarded the Nobel prize in 1906, and the Golgi apparatus, a structure in the cell body, was named after him. More than 100 years after Golgi's discovery, neurology continues to be one of the most exciting and progressive areas in the life sciences as still more secrets of the neuron are unraveled one by one.

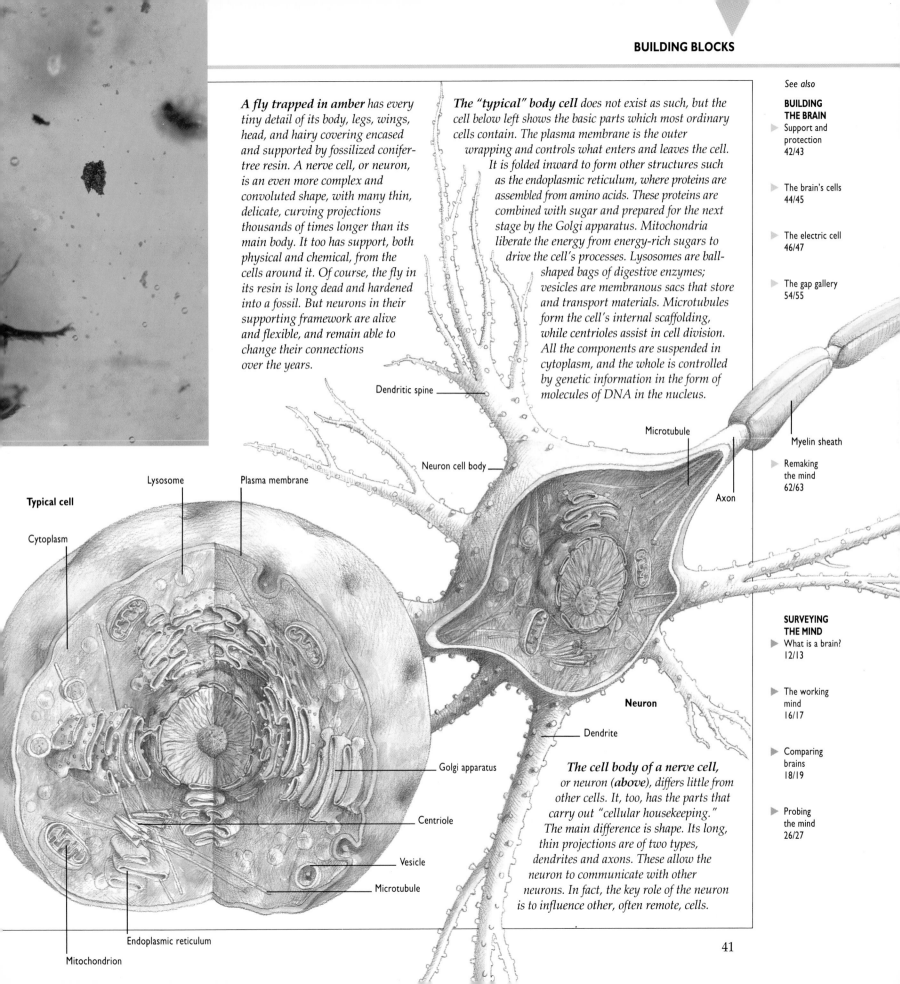

See also

**BUILDING
THE BRAIN**
▶ Support and
protection
42/43

▶ The brain's cells
44/45

▶ The electric cell
46/47

▶ The gap gallery
54/55

▶ Remaking
the mind
62/63

**SURVEYING
THE MIND**
▶ What is a brain?
12/13

▶ The working
mind
16/17

▶ Comparing
brains
18/19

▶ Probing
the mind
26/27

A fly trapped in amber has every
tiny detail of its body, legs, wings,
head, and hairy covering encased
and supported by fossilized conifer-
tree resin. A nerve cell, or neuron,
is an even more complex and
convoluted shape, with many thin,
delicate, curving projections
thousands of times longer than its
main body. It too has support, both
physical and chemical, from the
cells around it. Of course, the fly in
its resin is long dead and hardened
into a fossil. But neurons in their
supporting framework are alive
and flexible, and remain able to
change their connections
over the years.

The "typical" body cell does not exist as such, but the
cell below left shows the basic parts which most ordinary
cells contain. The plasma membrane is the outer
wrapping and controls what enters and leaves the cell.
It is folded inward to form other structures such
as the endoplasmic reticulum, where proteins are
assembled from amino acids. These proteins are
combined with sugar and prepared for the next
stage by the Golgi apparatus. Mitochondria
liberate the energy from energy-rich sugars to
drive the cell's processes. Lysosomes are ball-
shaped bags of digestive enzymes;
vesicles are membranous sacs that store
and transport materials. Microtubules
form the cell's internal scaffolding,
while centrioles assist in cell division.
All the components are suspended in
cytoplasm, and the whole is controlled
by genetic information in the form of
molecules of DNA in the nucleus.

Dendritic spine

Microtubule

Neuron cell body

Myelin sheath

Axon

Typical cell

Lysosome

Plasma membrane

Cytoplasm

Neuron

Dendrite

Golgi apparatus

Centriole

Vesicle

Microtubule

Endoplasmic reticulum

Mitochondrion

The cell body of a nerve cell,
or neuron (**above**), differs little from
other cells. It, too, has the parts that
carry out "cellular housekeeping."
The main difference is shape. Its long,
thin projections are of two types,
dendrites and axons. These allow the
neuron to communicate with other
neurons. In fact, the key role of the neuron
is to influence other, often remote, cells.

Support and protection

Evolution has provided a suitably secure system to cradle the delicate parts of that most precious organ, the brain.

The brain has the consistency of set yogurt. It is cradled within the rigid protective casket of the cranium, or "brain box." This box is formed by the firmly jointed curved bones of the skull. Layers of muscle, fat, skin, and (in most people) hair on the outside of the skull act as a shock-absorbing cushion for minor knocks and jars. Hair can also protect against extreme temperatures. Directly inside the skull bones are the three layers, or membranes, of the meninges, with cerebrospinal fluid (CSF) between the inner two. With this support, the brain "floats" within the skull and is shielded from shocks and vibrations.

At the microscopic level, the brain's neurons have a cellular scaffold of supporting cells, or neuroglia ("nerve glue"). Indeed, neuroglial cells form up to 90 percent of all cells in the brain and about half of the brain's total volume. Specialized capillary blood vessels create a "blood–brain barrier" – one that prevents invasion of the brain by potentially harmful chemicals from the blood. Finally, the brain is protected by a person's behavior: the head is moved quickly to duck a fast-moving ball, for example, or hazardous situations are avoided.

Cerebrospinal fluid (CSF) protects the brain in many ways. Outside, it occupies the subarachnoid space between the middle and inner meninges and provides hydraulic cushioning. Within, it flows slowly through a system of four interconnected branching cavities – the ventricles – supplying nutrients and removing wastes.

A network, or choroid plexus, of blood capillaries lining each ventricle produces CSF, which moves through the ventricles and into the subarachnoid space through openings in the fourth ventricle. Some flows down the spinal cord, but most seeps through arachnoid granulations to rejoin the blood in the superior sagittal sinus on top of the brain.

Third ventricle

Lateral ventricle

Fourth ventricle

Spinal canal

→ Circulation of CSF

→ CSF returning to blood

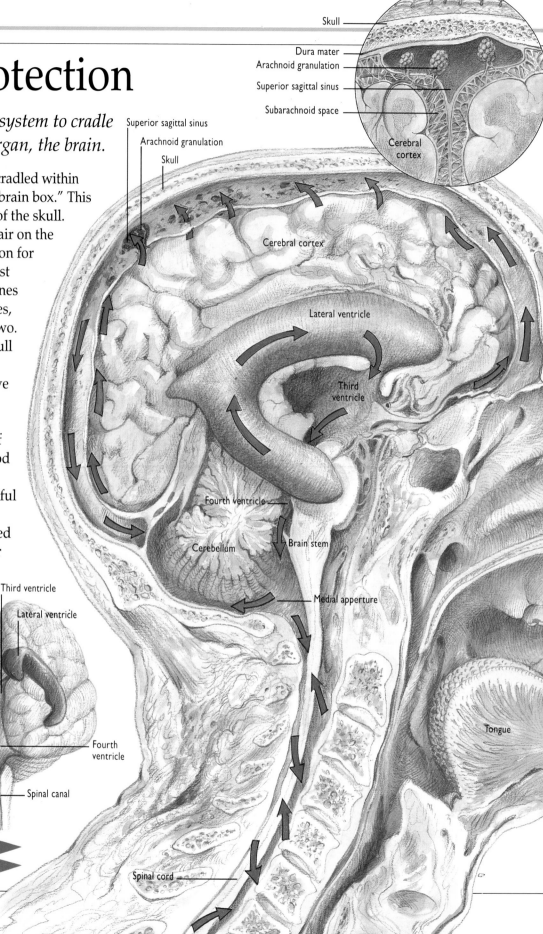

Skin

Skull

Dura mater

Arachnoid granulation

Superior sagittal sinus

Subarachnoid space

Cerebral cortex

Superior sagittal sinus

Arachnoid granulation

Skull

Cerebral cortex

Lateral ventricle

Third ventricle

Fourth ventricle

Cerebellum

Brain stem

Medial apperture

Tongue

Spinal cord

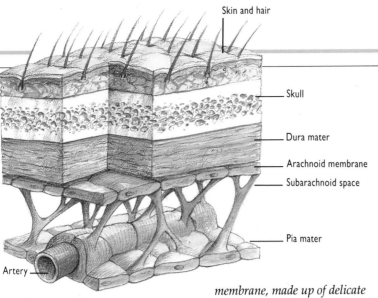

Skin and hair

Skull

Dura mater

Arachnoid membrane

Subarachnoid space

Pia mater

Artery

Nose

See also

**BUILDING
THE BRAIN**
▶ Building blocks
40/41

▶ The brain's cells
44/45

▶ The electric cell
46/47

▶ Sending signals
50/51

▶ Food for
thought
66/67

▶ The cell factory
72/73

**SURVEYING
THE MIND**
▶ The view
from outside
28/29

**INPUTS AND
OUTPUTS**
▶ Body links
80/81

▶ Survival sense
122/123

FAR HORIZONS
▶ The evolving
mind
126/127

A buffer of several layers shields the brain from the outside world. Hair and skin cover the hard bony skull and act as shock absorbers. The inside of the skull is lined by the dura mater, the outer of the three meninges, which is a relatively thick and tough membrane. The arachnoid membrane, made up of delicate connective tissue, separates the dura mater from the subarachnoid space, which contains vital cerebrospinal fluid. The thin innermost pia mater, rich in blood vessels that supply some of the blood to the brain, follows every contour of the brain's surface.

Debris-scavenging microglial cells are usually found between neurons and along blood vessels in the central nervous system. They are not true glial cells, but are part of the group of white blood cells known as macrophages — they patrol between neurons and neuroglia, searching for debris to engulf and destroy. Microglia are one of the four types of neuroglia, along with ependymal cells, astrocytes, and oligodendrocytes. The microglia above, in bright yellow, are from a cranial nerve of a rat.

Ependymal cells line the ventricles and the spinal canal. They are covered with fingerlike microvilli and allow certain molecules from the brain to pass into the CSF.

Ependymal cell

Star-shaped astrocytes insulate neurons, provide nutrients, and form their cellular supporting framework, especially during early development of the embryo. They also recycle neurotransmitter chemicals, and some can transmit their own electrical signals. With oligodendrocytes, astrocytes form the group of neuroglial cells called macroglia.

Protoplasmic astrocyte

Blood cell

Pericapillary end foot

Capillary

The blood–brain barrier protects the chemical environment of neurons and neuroglia in brain tissue, keeping it within precise limits. Blood vessels and capillaries in the brain are lined with a closeknit layer of cells that restrict the movement of substances between them. The cells themselves also have limited permeability, so only a few highly selected types of molecules can pass through them, from the blood into the brain.

Oligodendrocytes are neuroglial cells that make myelin in the brain. Spiral layers of this fatty substance insulate the axons of some neurons. (In the peripheral nervous system, Schwann cells do this task.)

Axon

Neuron

Oligodendrocyte

Subpial end foot

Fibrous astrocyte

Microglial cell

Myelin sheath

Perineuronal end foot

Pia mater

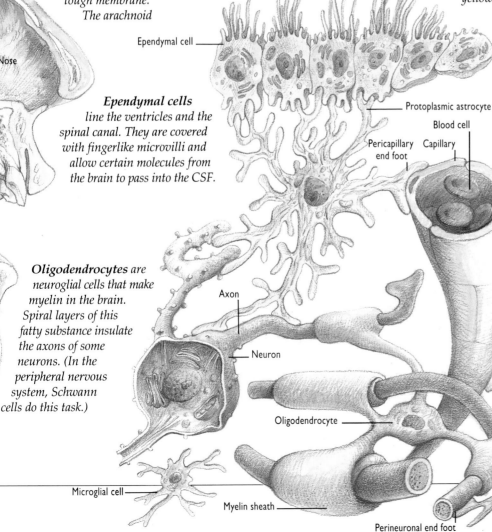

Synapse

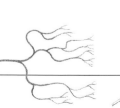

Myelin sheath

The brain's cells

Nerve cells come in an extraordinarily wide range of shapes and sizes.

Neurons, or nerve cells, are the basic units of the entire nervous system. There are probably about 100 billion neurons in the brain itself and at least the same number again in the rest of the nervous system. In its cell body, or soma, the typical neuron is much like any other cell. It is average in size, at 15–25 micrometers ($\frac{3}{5,000}$–$\frac{1}{1,000}$ inch) across, and has a cell nucleus, or control center, and other parts, or organelles, found in most types of cells.

So what makes the neuron special? Two main features. The first is that a neuron's outer layer – the plasma membrane – is specialized to convey nerve signals as electrochemical pulses. The second is its overall shape – the average neuron is not a sphere or cylinder like so many other cells. It has numerous projecting parts

known as neurites, which are like long tentacles or wires. They snake through the tissues to make connections with other neurons, so that nerve signals can pass between them. There are two main types of neurites: axons and dendrites. Axons are usually long, and each neuron has only one, which may branch; dendrites are shorter and have multiple branches.

In most neurons, especially those in the brain, the dendrites are so numerous and branching that they represent 90 percent or more of the neuron's total surface area. The geometry of its neurites determines the physical shape of a neuron and the connections it can make, which in turn determine its role and function.

Stellate neurons, found in the cerebral cortex and some regions of the brain stem and spinal cord, have neurites that occupy an approximately spherical volume. In the cortex their role is to deal with local processing, and they send messages to other cells that are in their immediate vicinity.

Stellate neurons

Golgi neuron

Pyramidal neuron

Bipolar neuron

Unipolar neuron

Different shapes of neurons *predominate in different parts of the nervous system. The simplest design is the unipolar neuron found mainly in the sensory systems. Its cell body (dark circle) does not take much part in signal conduction. The bipolar neuron collects signals from other cells via its dendrites, which all merge at the cell body, and passes its own signal along one main axon. The pyramidal neuron – the most common type in the cerebral cortex – has a roughly pyramid-shaped cell body. It has two widely separated sets of dendrites, as well as axon branches that may link back to them. The large Golgi neuron is found in the cerebellum, the part of the brain that has a role in fine movement.*

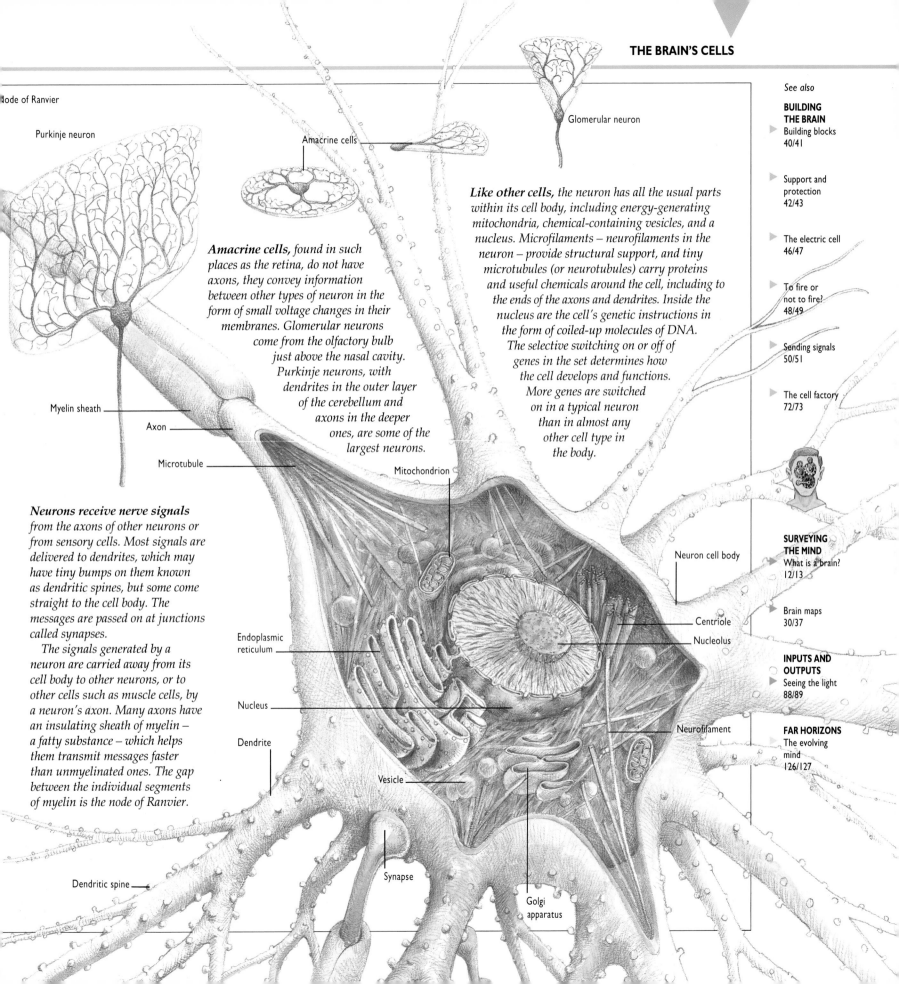

Node of Ranvier

Purkinje neuron

Amacrine cells

Glomerular neuron

Amacrine cells, found in such places as the retina, do not have axons, they convey information between other types of neuron in the form of small voltage changes in their membranes. Glomerular neurons come from the olfactory bulb just above the nasal cavity. Purkinje neurons, with dendrites in the outer layer of the cerebellum and axons in the deeper ones, are some of the largest neurons.

Like other cells, the neuron has all the usual parts within its cell body, including energy-generating mitochondria, chemical-containing vesicles, and a nucleus. Microfilaments – neurofilaments in the neuron – provide structural support, and tiny microtubules (or neurotubules) carry proteins and useful chemicals around the cell, including to the ends of the axons and dendrites. Inside the nucleus are the cell's genetic instructions in the form of coiled-up molecules of DNA. The selective switching on or off of genes in the set determines how the cell develops and functions. More genes are switched on in a typical neuron than in almost any other cell type in the body.

Myelin sheath

Axon

Microtubule

Mitochondrion

Neurons receive nerve signals from the axons of other neurons or from sensory cells. Most signals are delivered to dendrites, which may have tiny bumps on them known as dendritic spines, but some come straight to the cell body. The messages are passed on at junctions called synapses.

The signals generated by a neuron are carried away from its cell body to other neurons, or to other cells such as muscle cells, by a neuron's axon. Many axons have an insulating sheath of myelin – a fatty substance – which helps them transmit messages faster than unmyelinated ones. The gap between the individual segments of myelin is the node of Ranvier.

Endoplasmic reticulum

Nucleus

Dendrite

Neuron cell body

Centriole

Nucleolus

Neurofilament

Vesicle

Dendritic spine

Synapse

Golgi apparatus

See also
BUILDING THE BRAIN
Building blocks 40/41

Support and protection 42/43

The electric cell 46/47

To fire or not to fire? 48/49

Sending signals 50/51

The cell factory 72/73

SURVEYING THE MIND
What is a brain? 12/13

Brain maps 30/37

INPUTS AND OUTPUTS
Seeing the light 88/89

FAR HORIZONS
The evolving mind 126/127

The electric cell

Nerve cells have outer membranes with extraordinary properties vital to their ability to send messages.

Electricity is a natural part of living and is involved in every cell in the body. The myriad biochemical reactions and pathways that make up life processes take place in solution, with substances dissolved in water. When a substance such as common salt dissolves, it no longer exists as the usual atoms and molecules – in the case of salt, sodium chloride. The atoms lose or gain electrons to become free-floating particles known as ions. These have a positive or negative charge, and since electricity is the movement of charge, the movement of charged ions inside a living cell is electricity. In a typical cell, certain ions are kept in specific positions on each side of the cell membrane. But ions tend to drift around so that they become evenly spread out and balanced – they will diffuse until their concentration is equal, and any charge difference has been eliminated. Keeping ions separated against their tendency to equalize charge and concentration uses up energy, which the cell has to expend.

By means of a pump, a neuron's so-called excitable cell membrane is able to restrain ions from passing freely

via channels into or out of the cell. Thus, a neuron that is "on standby" keeps an excess of positive ions, or cations, just outside its cell membrane, and an excess of negative ions, or anions, inside. This separation of particles with opposite charges produces a potential difference, an electrical difference, between the inside and the outside of the membrane. It is known as the resting potential and can be measured with sensitive scientific equipment at around –70 millivolts, with the inside of the cell negatively charged.

When the neuron comes "off standby" and actually sends its signal, ions cross the membrane for a split second to make the inside momentarily positive by about 50 millivolts. This rise in voltage is an action potential. The shift in charge as the ions move is an electrochemical event and produces an electrical pulse – the nerve signal.

Sodium and chloride ions

An atom is made up of a positively charged central nucleus and particles called electrons which spin around it and which have an equivalent negative charge, so the whole atom is electrically neutral. In solution, an atom loses or gains one or more electrons. The result is a particle, or ion, that has charge. In crystals of common salt, sodium chloride (NaCl), the sodium and chlorine atoms are or bonded by sharing one electron. In

solution, they separate. Sodium loses the electron to form a positive ion, Na+; chlorine gains one to make a negative ion, Cl−. These two, along with the potassium ion K+, are the main ones involved in nerve impulses.

Sodium chloride (NaCl)

Positively charged sodium ion (Na+)

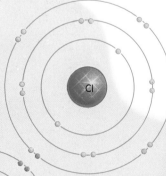

Negatively charged chloride ion (Cl−)

Water (H₂O)

An archer draws the bowstring and holds it steady, using muscle power to maintain enforced equilibrium. A small movement will release the bow's stored (or potential) energy. The neuron also uses energy to maintain the resting potential by separating charged particles across its cell membrane. Just a small

Inside the membrane, there is an overall excess of negative ions (anions); outside, an excess of positive ions (cations). Since opposite charges attract, positive and negative ions line up on each side of the membrane.

Axon

Membrane

Sodium
(Na$^+$)

Potassium
(K$^+$)

Chloride
(Cl$^-$)

Protein
anions

The ions on each side of the axon's cell membrane are dissolved in the fluid (mainly water) that is found both inside and outside all body cells.

change will cause the enforced equilibrium of the resting potential to give way as the particles move through the cell membrane to balance out the artificially maintained potential difference. This movement causes an action potential.

The cell membrane allows the cell to generate a resting potential. This is done using energy-consuming pumps in the membrane which make sure that ions stay segregated against their natural tendency to spread out and equalize their electrical charges and chemical concentrations. Sodium ions are pumped out of the cell, and potassium ions pumped back in.

A battery, or electric cell, is a device designed to make electricity by separating positive from negative charge – usually by separating negative electrons from the rest of their now-positive atoms. In an alkaline dry battery, zinc powder is dissolved in potassium hydroxide. When connected into a circuit, the zinc gives up electrons to the manganese dioxide. These flow around the conductor as an electric current. A living cell, a type of "biological battery," also separates charge across its membrane.

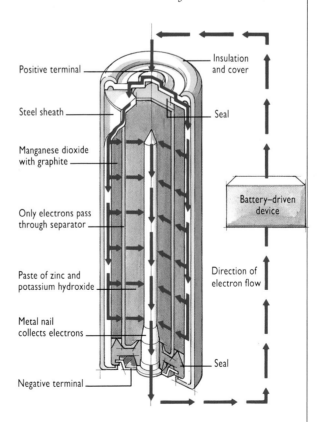

Positive terminal

Steel sheath

Manganese dioxide with graphite

Only electrons pass through separator

Paste of zinc and potassium hydroxide

Metal nail collects electrons

Negative terminal

Insulation and cover

Seal

Battery–driven device

Direction of electron flow

Seal

See also

BUILDING THE BRAIN
▶ Building blocks 40/41

▶ Support and protection 42/43

▶ The brain's cells 44/45

▶ To fire or not to fire? 48/49

▶ Sending signals 50/51

▶ Crossing the gap 52/53

▶ On or off 56/57

▶ Unlocking the gate 58/59

▶ Remaking the mind 62/63

▶ Food for thought 66/67

SURVEYING THE MIND
▶ Probing the mind 26/27

STATES OF MIND
▶ Anxious states 172/173

To fire or not to fire?

Every fraction of a second, each nerve cell in the brain and body decides whether or not to send a signal.

What prompts a nerve cell, or neuron, suddenly to fire a nerve signal, or action potential, along its axon? The answer depends largely on the signals that the neuron has itself received. These signals can be coming in at the rate of tens of thousands every second from other neurons which have connections (synapses) with the neuron in question. A neuron may have up to 100,000 synapses, tiny gaps crossed by signaling chemicals (neurotransmitters), which cause an electrical change when they arrive. This takes the form of a postsynaptic potential (PSP), a tiny strength change in the electrical potential – the difference in voltage – across the cell's membrane. Each PSP moves out from its synapse by a process called passive spread, fading with time and distance.

Every instant, waves of PSPs pass around the neuron's cell membrane. But they can trigger a signal only at the cone-shaped axon hillock where the axon joins onto the neuron's cell body. If the combined PSPs reaching the hillock push the potential there over a threshold level, an action potential of a fixed strength is fired, and this travels along the axon.

But an absence of inputs does not mean that a cell will not fire. Some neurons are autorhythmic – they can generate their own action potentials repeatedly (through oscillation in their membrane potential). Inputs to these neurons modulate their inherent rate of firing, turning it up or down. In other cases, a neuron will be quiescent until excited by an input. Information is coded in the brain and nervous system in the frequency of action potentials, the gaps between them, and the total numbers of neurons involved. For example, suppose a few neurons fire at a slow rate, and then a large number fire in rapid succession. If these two extremes represented signals from skin touch sensors, they might enable you to tell the difference between being hit by a ping-pong ball or a football.

Separate small explosive charges set at different places yet detonating together or in sequence can bring about the fall of a building. One small charge by itself would not have much effect. It is the adding, or summation, of the charges that works. A neuron works by a similar simple principle. Many small and variable events – incoming nerve signals – add together to tip the balance and trigger a large outcome, the outgoing nerve signal, or action potential.

When a neuron receives a signal from another neuron via one of its communication points, or synapses, neurotransmitter chemicals from the sending neuron diffuse across the synaptic gap and bind onto receptors in the membrane of the receiving neuron. These receptors either directly or indirectly cause tiny channels in the receiving cell's membrane to open, allowing ions (charged particles) into or out of the cell. This movement of ions creates a postsynaptic potential (PSP), which changes the potential of the neuron's membrane locally and generates a so-called electrotonic current which spreads passively through the neuron and the fluid surrounding it. The electrotonic current of a PSP travels to the axon hillock of the neuron, where it alters the potential across the membrane, which has an influence on whether or not the neuron sends its own action potential.

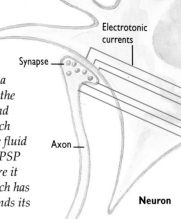

Electrotonic currents

Synapse

Axon

Neuron

Dendrite

Cell membrane

Inside the cell   Outside the cell

Inactivating particle

Activation gate

Voltage-gated ion channel

At the axon hillock, the potential across the cell membrane changes polarization according to the effects of all the various PSPs that take place in the neuron. PSPs can add, or combine, their effects if they occur closely spaced in time, a feature known as temporal summation. In the graph below, a sequence of PSPs – each one caused by the flow of ions through an ion channel, which takes a certain length of time – have occurred close enough together for the effects to accumulate.

The ensuing depolarization at the axon hillock has been sufficient to trigger an action potential. PSPs also add their effects if they are physically near one another on the neuron or its dendrites, a feature known as spatial summation. Any single PSP usually has to interact with many PSPs from other synapses. Some are not stimulatory, but inhibitory; the two types work to cancel each other out.

Sodium ion (Na⁺)

A single action potential is an all-or-nothing event of a fixed amplitude. It comes about when the tens of thousands of special ion channels clustered at a neuron's axon hillock open in concert. These are voltage-gated ion channels, and they stay closed until the local voltage (potential) at the hillock reaches the threshold – typically 10–20 millivolts depolarized from the resting potential of about –70 millivolts inside the membrane compared to outside. Temporarily, the potential changes by 120 millivolts to around +50 millivolts. Once triggered, the action potential proceeds at this strength along the length of the axon.

At rest, a voltage-gated ion channel is closed (**1**). When the polarization threshold is crossed, an activation gate on the inside of the channel swings open (**2**) and sodium ions flood inside (**3**), making the cell more positive. After a short while, a so-called inactivation particle swings shut, blocking the channel (**4**). When the polarization returns to normal, the activation gate resumes its closed position (**5**).

Action potential

Presynaptic impulses

Postsynaptic potential

Postsynaptic potential (millivolts)

–40

–60

Time

Axon hillock

Myelin sheath

See also

BUILDING THE BRAIN
▶ Building blocks 40/41

▶ The brain's cells 44/45

▶ The electric cell 46/47

▶ Sending signals 50/51

▶ Crossing the gap 52/53

▶ The gap gallery 54/55

▶ On or off 56/57

▶ Unlocking the gate 58/59

▶ Discovering transmitters 60/61

SURVEYING THE MIND
▶ Probing the mind 26/27

HOMING IN ON THE NERVE CELL'S SECRETS

Since 1976, researchers have had the means to study those parts of the nerve cell that are responsible for its remarkable characteristics. The technique that has made this possible – patch clamping – has allowed them to isolate and study single ion channels in the cell membranes of neurons. Indeed, this technique has proved so valuable that its main developers – Erwin Neher and Bert Sakmann – received the Nobel prize for medicine in 1991.

An incredibly fine glass pipette (tube) is placed on the cleaned membrane of a cell, and slight suction is applied up it. This seals the pipette's mouth, which may be 50,000 times finer than a human hair, against a small area, or patch, of membrane. The patch is isolated physically, electrically, and chemically. Wires in the pipette and in other parts of the cell either stimulate it, or measure electrical changes that result as ion channels in the isolated patch open and close. Various chemicals can also be introduced via fluid in the pipette, and their effects followed. Or the patch of membrane can be torn away from the cell for further study in isolation.

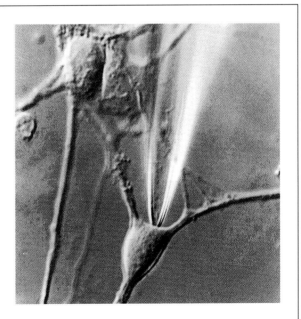

Sending signals

Tiny fibers carry signals from one nerve cell to another. Some are carried quickly, others more slowly.

When a neuron, or nerve cell, sends a message down its wirelike axon to another neuron, the signal moves along as a traveling electric pulse. The pulse is carried along by the sequential opening and closing of tiny pores in the axon's membrane. These pores, known as voltage-gated ion channels, respond to voltage changes, or depolarization. Below a certain threshold voltage, they stay closed; above that level, they open to allow charged particles (ions) to flow through them.

Since a flow of charged particles is, in effect, an electric current, this flow pushes the voltage of the axon membrane locally over the threshold and causes channels next to the ones that have just opened to open in turn. Thus a pulse, or action potential, travels along the axon.

Many axons are insulated with a spiral wrapping of myelin, a fat-rich substance formed by a special type of cell. Typically, each segment of myelin sheath is about 1/25 inch (1 mm) long, and it may have up to 100 layers. The myelin segments are arranged along the axon like a string of sausages. But they

*A nerve signal starts in the axon hillock of a nerve cell (**below left**). Here, negative inhibitory potentials and positive excitatory ones from elsewhere in the nerve cell coincide, and when the potential rises above a certain threshold, it causes voltage-gated ion channels in the initial segment of the axon to open. This sends a wave of depolarization, or an action potential, along the axon.*

The runners in a relay race do not complete the course, but the baton does. Each runner passes it in turn to the next runner in the sequence, for energy input and a fresh burst of speed. Similarly, a nerve signal jumps along the insulated sections of a myelinated axon, receiving a fresh burst of energy at each uninsulated section to speed it along.

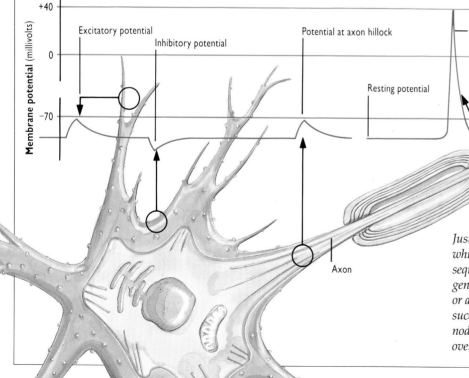

In a myelinated, or insulated, axon nerve signals do not travel smoothly. Just like a row of lights, each of which flashes momentarily and in sequence, a myelinated axon generates a series of nerve signals, or action potentials, in fast succession, which move from one node of Ranvier to the next. The overall effect is that the nerve signal

flashes along the axon. Because of this, myelinated axons conduct their signals faster and more efficiently than neurons whose axons lack a myelin sheath. Signals move along myelinated axons at speeds of 13–395 feet/sec (4–120 m/sec) — the thicker the axon, the faster the signal. In non-myelinated axons, speeds are much slower— 20 inches–13 feet/sec (0.5–4 m/sec).

See also

**BUILDING
THE BRAIN**
▶ Building blocks
40/41

▶ Support and
protection
42/43

▶ The brain's cells
44/45

▶ The electric cell
46/47

▶ To fire or
not to fire?
48/49

▶ Crossing
the gap
52/53

▶ Remaking
the mind
62/63

▶ Recovering
from damage
64/65

▶ Linking up
74/75

**SURVEYING
THE MIND**
▶ Discovering
through damage
24/25

**INPUTS AND
OUTPUTS**
▶ Body links
80/81

▶ The long
junction
82/83

MULTIPLE SCLEROSIS

In a myelinated axon – the wirelike extension of a neuron (nerve cell) that carries messages from the cell body – the fatty myelin sheath is a good electrical insulator. It prevents the electrochemical charges of a nerve signal from leaking away into the cells and fluid around the axon and conducts signals rapidly and energy efficiently.

Some medical conditions are caused by problems with the myelin sheaths, and this can have a great impact on the way that neurons communicate. One of these is multiple sclerosis, or MS, in which myelin is progressively destroyed, possibly due to the body's immune defense system mistakenly attacking itself. The demyelinating sites become inflamed and affected by non-functioning scar tissue (sclerosis and plaque formation).

This prevents the damaged neurons from sending messages. Symptoms generally develop slowly and depend on the nerves affected. They may include numbness or tingling in the hands and feet, visual disturbances, muscle weakness, slow movements and clumsiness, bladder and bowel problems, and personality changes. The symptoms tend to come and go, but in general, they progress over years. There is no cure for MS – treatment focuses on symptom relief, and helps for daily life.

MS usually begins around 20 to 25 years, but it can develop at any age from 10 to 50. It is relatively common in certain (mainly white) ethnic groups, affecting as many as 1 in 350 people. Yet it is almost unknown in other groups, such as some black Africans and Inuit Eskimos.

Layers of myelin

Axon

are not quite in contact with each other – there is a small gap between one segment of myelin sheath and the next. This gap is known as a node of Ranvier, and it acts like a "booster station" to the nerve signal, just as booster stations at regular intervals along a telephone cable amplify and resend fading signals.

Between these nodes, the myelin coating prevents contact between the axon membrane and the extracellular fluid, which contains the various ions needed for changing the voltage. This means that the depolarizing effects of the action potential spread passively along the membrane itself and inside it until a node of Ranvier is reached. Here the axon's membrane has extracellular fluid and ions on the outside and is very excitable, so the full action potential can be reconstituted and refired, making it appear to leap from node to node along the axon. It is known as saltatory conduction (from the Latin *saltare*, to jump). Axons that do not have a myelin sheath conduct their signals more slowly than insulated myelinated axons.

As a nerve signal, or action potential, travels along an axon's membrane, electrically charged sodium ions (Na⁺) rush to the inside of the membrane through voltage-gated channels. This temporarily changes the difference in voltage between the inside and the outside of the membrane from its resting potential of -70 mV to as much as +40 mV. This change is known as depolarization, and it makes adjacent Na⁺ channels open and a wave of depolarization travels along it. After Na⁺ floods in, potassium ions (K⁺) move out as the cell repolarizes. To restore the resting potential, pumps in the membrane remove the Na⁺ and import K⁺.

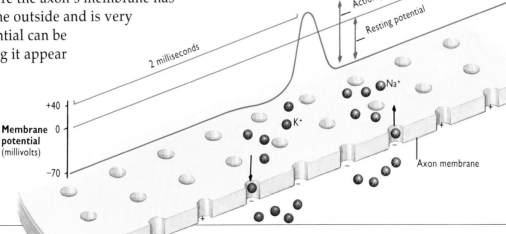

Direction of impulse

Action potential

Resting potential

2 milliseconds

+40

Membrane 0
potential
(millivolts)

−70

Na⁺

K⁺

Axon membrane

Ion channel

Crossing the gap

For one nerve cell to communicate with another, it has to transmit its signal across a small gap.

When a nerve signal, or impulse, reaches the end of its axon, there is a fundamental change in its nature. Along the axon, the nerve signal has traveled as an action potential – a pulse of electricity. But between one neuron and the next, there is no cellular continuity; there is a gap called the synapse. The membranes of the sending and receiving cells are separated by the fluid-filled synaptic gap.

Although the gap is only 20–25 nanometers across ($\frac{1}{600}$ the width of a hair), the signal cannot leap it electrically. So chemicals, or neurotransmitters, are released by the presynaptic "sending" membrane and seep across the gap to receptors on the receiving neuron's postsynaptic membrane. The binding of neurotransmitters to these receptors has the effect of allowing ions (charged particles) to pass in and out of the receiving cell. This amounts to a transient change in potential difference (that is, an electrical signal). This signal is not large enough on its own to be a full action potential, but contributes to the possible initiation of one at the axon hillock, the start of the receiving neuron's axon.

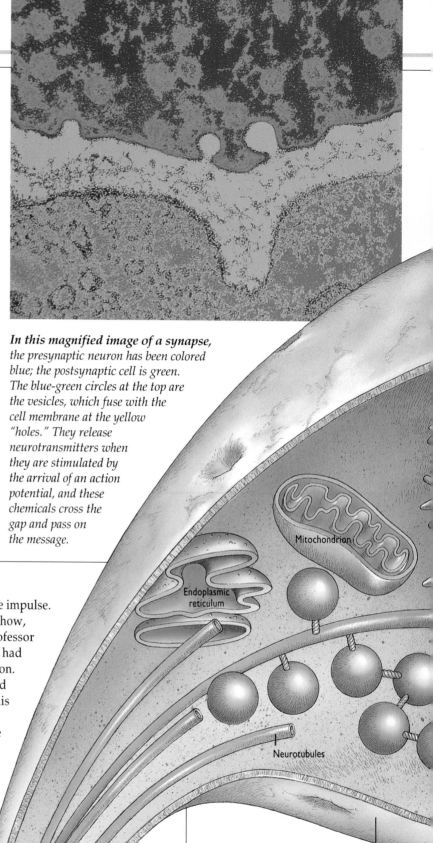

In this magnified image of a synapse, the presynaptic neuron has been colored blue; the postsynaptic cell is green. The blue-green circles at the top are the vesicles, which fuse with the cell membrane at the yellow "holes." They release neurotransmitters when they are stimulated by the arrival of an action potential, and these chemicals cross the gap and pass on the message.

Mitochondrion

Endoplasmic reticulum

Neurotubules

Axon

Presynaptic membrane

A REWARDING EASTER

In the early 1900s, much research was carried out into the nature of the nerve impulse. Some scientists suspected that it was not solely an electrical process; somehow, chemicals were involved. In 1921 physiologist Otto Loewi (1873–1961), professor of pharmacology in Graz, Austria, woke in the night of Easter Sunday. He had an idea to demonstrate chemical involvement in nerve-impulse transmission. He scribbled some notes and slept on, but in the morning he could not read his handwriting! The next night he woke again and immediately went to his laboratory and did the experiment. He used dissected frog hearts with the vagus nerve, which controls heart rate, still attached. Loewi immersed one frog heart in a fluid, Ringer's solution, and stimulated its vagus – which would normally slow the heartbeat. As expected, the beating rate reduced. When a second heart was put in the solution, its beat also decreased – without stimulation. Loewi reasoned that a chemical from the first heart had seeped through the solution and affected the beating of the second heart. After two hours, he had shown for the first time that chemicals – now called neurotransmitters – were involved in nerve-impulse transmission. In 1933 the chemical involved, acetylcholine, was isolated. Dozens of neurotransmitters have since been discovered.

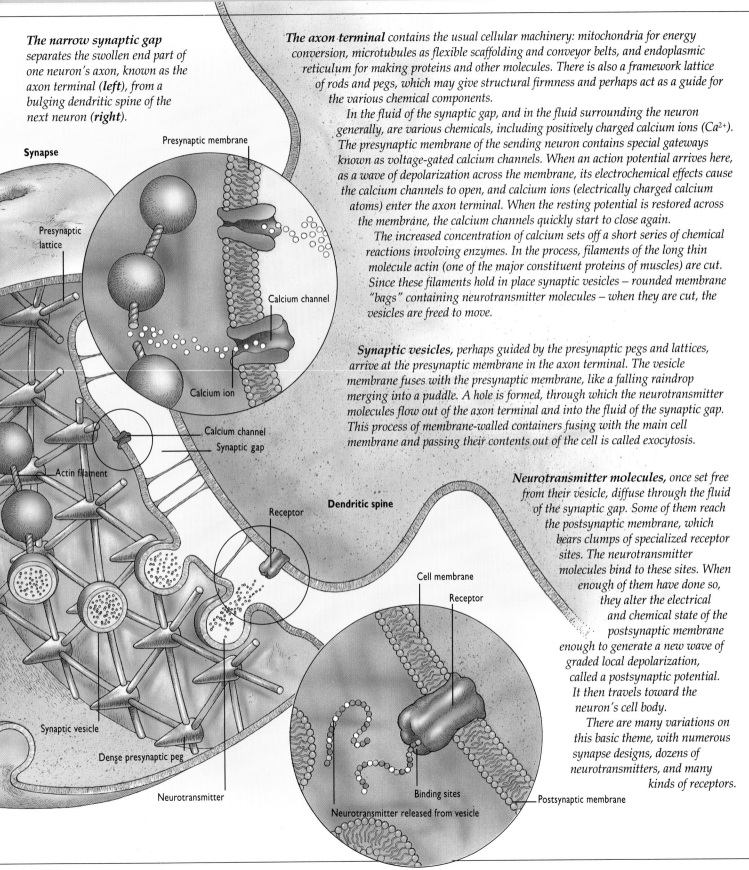

The narrow synaptic gap *separates the swollen end part of one neuron's axon, known as the axon terminal (left), from a bulging dendritic spine of the next neuron (right).*

Synapse

Presynaptic membrane

Presynaptic lattice

Calcium channel

Calcium ion

Calcium channel

Synaptic gap

Actin filament

Receptor

Dendritic spine

Cell membrane

Receptor

Synaptic vesicle

Dense presynaptic peg

Neurotransmitter

Neurotransmitter released from vesicle

Binding sites

Postsynaptic membrane

The axon terminal *contains the usual cellular machinery: mitochondria for energy conversion, microtubules as flexible scaffolding and conveyor belts, and endoplasmic reticulum for making proteins and other molecules. There is also a framework lattice of rods and pegs, which may give structural firmness and perhaps act as a guide for the various chemical components.*

In the fluid of the synaptic gap, and in the fluid surrounding the neuron generally, are various chemicals, including positively charged calcium ions (Ca^{2+}). The presynaptic membrane of the sending neuron contains special gateways known as voltage-gated calcium channels. When an action potential arrives here, as a wave of depolarization across the membrane, its electrochemical effects cause the calcium channels to open, and calcium ions (electrically charged calcium atoms) enter the axon terminal. When the resting potential is restored across the membrane, the calcium channels quickly start to close again.

The increased concentration of calcium sets off a short series of chemical reactions involving enzymes. In the process, filaments of the long thin molecule actin (one of the major constituent proteins of muscles) are cut. Since these filaments hold in place synaptic vesicles – rounded membrane "bags" containing neurotransmitter molecules – when they are cut, the vesicles are freed to move.

Synaptic vesicles, *perhaps guided by the presynaptic pegs and lattices, arrive at the presynaptic membrane in the axon terminal. The vesicle membrane fuses with the presynaptic membrane, like a falling raindrop merging into a puddle. A hole is formed, through which the neurotransmitter molecules flow out of the axon terminal and into the fluid of the synaptic gap. This process of membrane-walled containers fusing with the main cell membrane and passing their contents out of the cell is called exocytosis.*

Neurotransmitter molecules, *once set free from their vesicle, diffuse through the fluid of the synaptic gap. Some of them reach the postsynaptic membrane, which bears clumps of specialized receptor sites. The neurotransmitter molecules bind to these sites. When enough of them have done so, they alter the electrical and chemical state of the postsynaptic membrane enough to generate a new wave of graded local depolarization, called a postsynaptic potential. It then travels toward the neuron's cell body.*

There are many variations on this basic theme, with numerous synapse designs, dozens of neurotransmitters, and many kinds of receptors.

See also
BUILDING THE BRAIN
▶ The electric cell 46/47

▶ Sending signals 50/51

▶ The gap gallery 54/55

▶ Unlocking the gate 58/59

▶ Discovering transmitters 60/61

▶ Remaking the mind 62/63

▶ Recovering from damage 64/65

INPUTS AND OUTPUTS
▶ The long junction 82/83

▶ On the move 84/85

STATES OF MIND
▶ Physical cures 166/167

▶ The mind adrift 170/171

▶ Anxious states 172/173

The gap gallery

Synapses – their design, structure, and position – are the key to how nerve signals are transmitted.

Between two nerve cells (neurons), one sending and one receiving a signal, there is a synapse, or synaptic gap, that can take many forms. The synapse's sending part is usually the terminal of an axon, the signal-carrying "wire" of a neuron. The receiving part may be on the shaft of a neuron's dendrite; on the cell body – or soma – of the neuron; or on a bulge, or "spine," projecting from the dendrite or the soma.

When a signal is received, the voltage in the membrane of the receiving neuron changes the polarization locally. This passes like a current in a wire to the axon hillock, a part of the receiving cell close to its body. Each millisecond, a neuron is subjected to waves of polarization changes coming in from up to 100,000 synapses. If these changes are strong enough and/or occur close enough together to take the voltage at the axon hillock over a threshold level, the receiving cell in turn sends a signal and transmits a message down its own axon.

An axosomatic synapse (far left) has a powerful effect because it is so close to the nerve cell body. The closer a synapse is to the cell body, the greater its effect. So if the two axodendritic synapses below sent a signal simultaneously, the one on the left would have more influence on the receiving cell.

In most large companies, decision-making depends on many sources of information. Some input – like that of department heads or board members – carries more weight. But some influence is exerted at all levels, from staff on the factory floor to those in purchasing. On the way to the boardroom, where final corporate decisions are made, ideas and proposals are filtered, coordinated, added, and integrated.

Similarly, the neuron is subjected to myriad influences when it comes

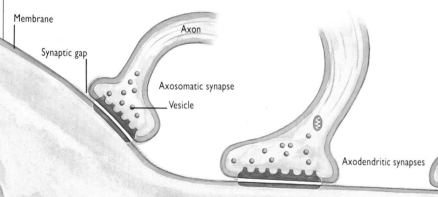

Membrane

Synaptic gap

Axon

Axosomatic synapse

Vesicle

Axodendritic synapses

Cell body

Dendrite

Dendritic spine

Axospinodendritic synapse

Synapses are named after their component parts. Thus, the synapse left is termed axospinosomatic since it has an axon from a sending neuron on one side of its gap and a spine projecting from a receiving nerve cell body, or soma, on the other side.

At an axospinodendritic synapse (right), an axon forms a synapse with a spine on the dendrite of a receiving neuron. They are the most common, forming 60 percent of all synapses in the human brain. Axodendritic synapses (above) make up 30 percent, axosomatic ones 6 percent.

See also

BUILDING THE BRAIN
▶ The brain's cells 44/45

▶ To fire or not to fire? 48/49

▶ Sending signals 50/51

▶ Crossing the gap 52/53

▶ On or off 56/57

▶ Remaking the mind 62/63

▶ Linking up 74/75

INPUTS AND OUTPUTS
▶ Body links 80/81

STATES OF MIND
▶ Physical cures 166/167

A STAGGERING POTENTIAL

The human brain has about 100 billion nerve cells, or neurons. Each neuron can have tens of thousands of links with other neurons. The number of possible routes for nerve signals through this vast maze defies contemplation. The permutations are greater in number than all the fundamental particles – electrons, protons, neutrons, and so on – in the universe. Our brain's complexity, plus limited study opportunities, means that scientists often look to smaller, simpler creatures to try to discover its principles of operation.

A grasshopper has around 16,000 neurons in the main nerve ganglia that function as its "brain." This is a fraction of the number of human neurons.

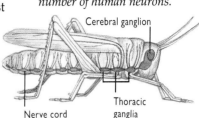

Cerebral ganglion

Thoracic ganglia

Nerve cord

to its "corporate decision": whether to fire its own signals, or action potentials. Potentials of different strengths arrive at the cell body from the thousands of synapses on its dendrites. Synapses that make important contributions are generally those closest to the neuron's main cell body. Potentials from the more remote dendrites have faded by the time they reach the cell body, just like the decisions in the more remote parts of a company, far from headquarters.

There are two main types of synapses, *excitatory and inhibitory. Excitatory synapses generate polarization changes that are more likely to make the cell fire (send a signal); inhibitory ones are less likely to make the cell fire. In the axospinodendritic synapses below, the inhibitory synapse counteracts the excitatory. If both axons try to send a signal across the synapse at the same time, there may be no overall change. If one axon becomes active while the other is quiet, its influence is passed on.*

In an axoaxonic synapse (below), *the axon from one neuron synapses with an axon from a second neuron; this second axon also synapses with a third neuron, for instance on a dendritic spine. The first axon can modulate the action of the second – if it has an inhibitory effect, it might stop the second axon from passing its message to the third neuron. This is known as presynaptic inhibition. In presynaptic excitation, the first axon works with the second to create more excitatory activity in the third neuron. At an electrical synapse, the axon and dendrite actually touch, with no gap between them.*

Axoaxonic synapse

Axon

Axospinodendritic synapses

Inhibitory synapse

Excitatory synapse

Electrical synapse

Synapse

Bulge in axon

Axon

A synapse *does not always occur at the end of an axon; it can, in fact, form part of the way along the axon's length (left). This can only happen where the axon is not covered by an insulating myelin sheath. Typically, the axon bulges at this type of synapse.*

Dendrodendritic synapse

At a dendrodendritic synapse (above), *a dendrite forms a functional contact with another dendrite. This type of synapse is relatively rare.*

Dendrite

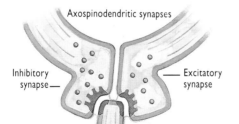

On or off

A nerve cell's rate of signaling is governed by a complex control system with competing on–off inputs.

The effect of a nerve signal after it crosses the synapse – the gap across which messages pass from one neuron to the next – is vital to the workings of the brain and nerves. Along the axon of a sending neuron, it travels as an action potential, a wave of depolarization. After crossing the synapse in chemical form, it is not big enough to constitute an action potential, but continues in the receiving neuron as a smaller depolarization, a postsynaptic potential (PSP), which fades with distance and time. This signal is integrated with other PSPs from other axons.

The neurotransmitter that crosses the synaptic gap and the target receptor that binds it govern what happens in the receiving cell. Some neurotransmitters pass across the synapse to receptors and cause depolarization. This depolarization excites the receiving neuron, so it is more likely to fire its own action potentials. It is known as an excitatory postsynaptic potential (EPSP), the "on" side of the system.

Other neurotransmitters produce inhibition – the "off" side of the system. This involves not depolarization but hyperpolarization. The inside of the postsynaptic membrane briefly becomes more negative than normal with respect to the outside. This inhibitory postsynaptic potential (IPSP) dampen down the receiving neuron by working against the effects of EPSPs and makes it less likely that the neuron will fire its own action potentials. Still other neurotransmitters can produce either inhibition or excitation depending on the receptors that are waiting for them. One of the principal inhibitory ones, gamma-aminobutyric acid (GABA), is found at about one-third of all brain synapses and inhibits neuronal firing in such parts as the brain stem, cerebellum, and cerebral cortex.

The neuron integrates both "on" and "off" inputs. On signals are received from synapses where an excitatory postsynaptic potential (EPSP) is produced, and off signals come from synapses where an inhibitory postsynaptic potential (IPSP) is made. These potentials move as currents to the axon hillock of the cell. The EPSPs cause depolarization, the IPSPs hyperpolarization, and these two conflicting currents tend to cancel each other out.

The final addition and subtraction of depolarizations and hyperpolarizations (and there are thousands of each from a cell's many synapses) takes place at the neuron's axon hillock. If the potential here reaches its threshold, the neuron is excited enough to fire its own action potential, sending a signal along its axon.

Dendrite

Excitatory synapse

Inhibitory synapse

Excitatory synaptic current

Inhibitory synaptic current

Axon hillock

Depolarization

Axon

THE ELECTRICAL STORM

Once seen as a form of supernatural possession, epilepsy is now recognized as a malfunction of brain neurons. It appears to have various causes and involve different parts of the brain, so each type has distinct symptoms. Grand mal epilepsy affects motor areas, causing the jerky, spasmodic movements of a seizure. The petit mal form involves sensory brain areas, making sufferers experience strange sights, sounds, smells, and thoughts. Symptomatic epilepsy is due to an underlying condition such as a brain abscess or tumor. Idiopathic epilepsy has no obvious cause.

EEG traces of epilepsy sufferers show a mass of random signals that have temporarily lost their natural peaks, valleys, and rhythms. The cells seem to be superexcited, with no inhibitions, and are set off by the slightest event. This may affect just one area of the brain, a partial seizure, or spread through the brain as a generalized seizure.

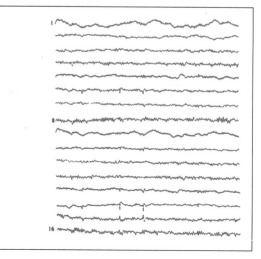

See also

**BUILDING
THE BRAIN**

▶ The electric cell
46/47

▶ To fire or
not to fire?
48/49

▶ Sending signals
50/51

▶ Crossing
the gap
52/53

▶ The gap gallery
54/55

▶ Unlocking
the gate
58/59

▶ Discovering
transmitters
60/61

▶ Remaking
the mind
62/63

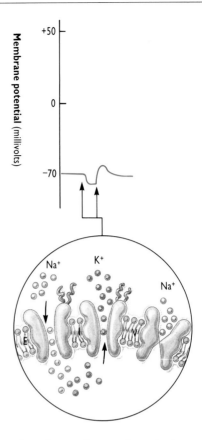

Membrane potential (millivolts)

+50

0

−70

Action potential

Membrane potential (millivolts)

+50

0

−70

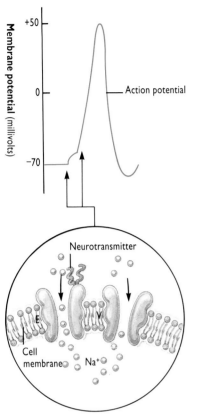

Neurotransmitter

Cell membrane   Na⁺

E —Excitatory receptor

I —Inhibitory receptor

V —Voltage-gated ion channel

Excitation is produced by EPSPs. Neurotransmitters cause an excitatory receptor (E) to allow sodium ions (Na^+) into the cell across the postsynaptic membrane, creating a depolarizing postsynaptic potential that passes to the axon hillock. Here, if it reaches its threshold level, the voltage-gated ion channel (V) opens, allowing sodium ions to flood in, triggering an action potential (**left**). Usually it takes a number of EPSPs to produce the waves of depolarization that make the neuron fire its own action potential.

Inhibition is caused by IPSPs. An IPSP arises when neurotransmitters make an inhibitory receptor (I) open its ion channel. The inhibitory receptor shown here allows positive potassium ions (K^+) to leave the cell. Other typical inhibitory receptors allow negative chloride ions (Cl^-) to enter the cell. The effect in each case is to cause hyperpolarization – the inside of the cell becomes more negative with respect to the outside. This counteracts the effect of EPSPs, making firing less likely, since the necessary threshold level will not be reached, and so the voltage-gated channel will not open (**right**).

Na⁺   K⁺   Na⁺

A racing car has two important foot-pedal controls – accelerator for "go" and brake for "stop." Without brakes, the car is far more difficult to control. The driver can only coast and slow down by letting natural resistance such as friction with the air and in the engine and gears take its effect. Compared to braking, this is a more lengthy and less precise process.

Neurons in the brain also have accelerators and brakes. Excitatory synapses press the accelerator and encourage the receiving neuron to fire its own signals, and fire them more rapidly. Inhibitory synapses apply the brakes and have a slowdown effect, adding another layer of control and precision to the overall functioning.

**INPUTS AND
OUTPUTS**

▶ On the move
84/85

▶ Contrasting
views
90/91

▶ On the scent
108/109

**STATES OF
MIND**

▶ Physical cures
166/167

Unlocking the gate

Like locks opened by chemical keys, tiny receptors are essential parts in the transmission of nerve signals.

Neurotransmitters are the molecules responsible for the chemical transmission of nerve signals across synapses. They act like keys that fit into the locks of receptors – protuberances of protein in the cell membrane of a neuron. When a receptor accepts a specific neurotransmitter, it either directly or indirectly opens channels through which ions (charged particles) flow. This has the effect of causing electrical changes in the membrane of the cell receiving the signal, which result in a postsynaptic potential (PSP).

There may be as many individual types of receptors as there are neurotransmitters – more than 60. Our knowledge of receptors has lagged behind that of neurotransmitters, but recent research is revealing their secrets. It seems that in the human brain and spinal cord, receptors fall into two broad groups. These are the neurotransmitter-gated ion channels (NGICs) and the G-protein linked receptors (GPLRs).

Detail of M2 segment

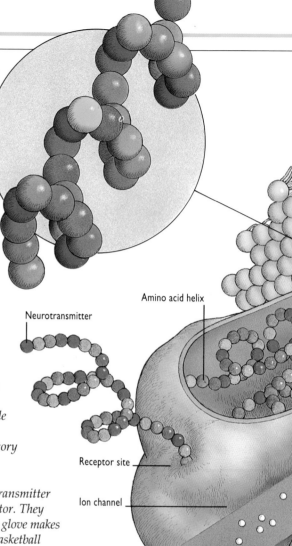

Neurotransmitter molecules activate a neurotransmitter-gated ion channel (NGIC) when they bind onto receptor sites on sub-units of the NGIC (four of which can be seen as bulges in the large diagram). This alters its shape, opening a channel in the middle.

The tunnel is lined by five coiled M2 segments (see below). The shapes and electrical charges on these segments draw ions through the channel into the cell. In this NGIC, the ions are positive, making the cell more positive inside and so contributing to its depolarization – an excitatory postsynaptic potential.

Amino acid helix

Neurotransmitter

Receptor site

Ion channel

Like a baseball in a baseball glove, a neurotransmitter molecule fits precisely with its intended receptor. They are truly made for each other. The shape of the glove makes catching the correct ball easier and surer. A basketball would not fit in the glove at all, and a golf ball would rattle around loosely. Likewise, the characteristics of a neurotransmitter molecule – its size, shape, electrical charges, bonding sites, and other features – are exactly fitted to a certain receptor or receptor group and no others.

A neurotransmitter-gated ion channel (NGIC) is effectively one large doughnut-shaped protein structure, 12 nanometers ($\frac{1}{2,000,000}$ inch) long. It is set in the "receiving" part of the synapse, the postsynaptic cell membrane of the neuron. Most of it projects out of the cell membrane, into the synaptic gap where it can come into contact with neurotransmitters.

The NGIC is made of five sub-units, each of which consists of four coils, or helixes, of amino acids (building blocks of proteins). The coils are named M1 to M4. Five M2 helixes – one for each sub-unit – form the center, or "ring," of the doughnut. They control the flow of ions through the NGIC.

To activate a postsynaptic receptor, an action potential arrives at the presynaptic membrane and causes the release of neurotransmitters, which drift across the synaptic gap. In an NGIC, the neurotransmitter-receiving sites and ion channel are part of the same structure, so this type of receptor works relatively fast.

Section through neurotransmitter-gated ion channel

Ion

The first type of NGIC to be studied was the nicotinic receptor for the neurotransmitter acetylcholine – in a type of fish, the electric ray. Similar types of these doughnut-shaped receptors have now been found in other animals and humans. The various designs of NGICs work with other neurotransmitters such as GABA, glycine, glutamate, and serotonin.

Since some of these chemicals are excitatory and others inhibitory, the design of the ion channel must be adapted to suit both positive ions, such as sodium and potassium, and negative ions such as chloride. GPLRs work in a different, more complex, way, using a "second messenger" system. They are also more widespread and variable than NGICs. They include receptors for norepinephrine, dopamine, and serotonin.

Cell membrane

REMOTE CONTROL

The G-protein linked receptor (GPLR) has an indirect action on ion channels. It has its effect through a series of biochemical steps using the second messenger molecule known as G-protein (short for GTP-binding protein). A typical GPLR is a single protein molecule in the form of a long, accordion-shaped chain. Its sections cross through the cell membrane a characteristic seven times as though it has been stitched in place. When a molecule of the relevant neurotransmitter arrives, it binds to the GPLR (green) and changes the receptor's shape. This has the effect of attracting the G-protein complex (yellow) to it. The alpha sub-unit, one of several sub-units in the G-protein, moves to a target in the cell body or membrane, where it triggers more reactions. When the GPLR loses its neurotransmitter, the alpha sub-unit joins the G-protein again, which then detaches from the GPLR.

G-protein complex

G-protein linked receptor

Alpha sub-unit

Cell membrane

Target

When an ion channel is the target of a G-protein, the net effect – the opening of the channel to allow ions in or out – is much the same as for an NGIC, but as the system is more long-winded, it takes longer. There are various types of G-protein; they can have excitatory or inhibitory effects on different targets; and different G-proteins set free by different neurotransmitters may even compete for the same target.

G-proteins can home in on enzymes as well as ion channels. An enzyme (the green coil) activated by a G-protein can bring about still more reactions, producing new chemicals. These might not necessarily be involved in the opening or closing of ion channels in the receiving cell's membrane. For instance, the G-protein known as Gq leads – via a cascade of reactions – to changes in protein levels in the cell.

See also

BUILDING THE BRAIN
▶ The electric cell 46/47

▶ To fire or not to fire? 48/49

▶ Crossing the gap 52/53

▶ On or off 56/57

▶ Discovering transmitters 60/61

▶ Remaking the mind 62/63

SURVEYING THE MIND
▶ Probing the mind 26/27

INPUTS AND OUTPUTS
▶ Seeing the light 88/89

STATES OF MIND
▶ Physical cures 166/167

▶ Anxious states 172/173

Discovering transmitters

Nerve-signal transmission between neurons – using messenger chemicals called neurotransmitters – is exquisitely varied and complex.

In order to jump the gap between a sending, or presynaptic, neuron and a receiving, or postsynaptic, neuron, a nerve signal makes use of chemicals – neurotransmitters. There are at least 60 different ones, and more are being discovered all the time. Some are generally excitatory, making receiving neurons more likely to fire; others are inhibitory, damping down their activity. Some can be either, depending on the part of the nervous system involved, the receptors waiting for them, and the ways that these receptors are linked into the machinery of the postsynaptic membrane. Different neurotransmitters may activate receptors which are all "wired" to one type of ion channel, so they cause one type of electrochemical response. This is known as convergence. Or, the same neurotransmitters can stimulate receptors that are connected to different ion channels, producing several responses, or divergence.

Researchers are finding ever greater levels of sophistication, and their work is having immense practical consequences, too. This is because the chemistry of neurotransmitter-receptor systems is one of the few stages in the process of brain and nerve function that is amenable to outside manipulation – using drugs. Some drugs work against neurotransmitters by disrupting their production or release, or by attaching to them as they cross the synaptic gap, so that they cannot fit into their tailor-made receptors. Others mimic

*When **neurotransmitters are released** from their vesicles at the presynaptic membrane in response to the arrival of an action potential, they enter the synaptic gap. Some diffuse across the gap and reach their target receptors, but many do not. Whether they hit their target or not, they have to be cleared away or broken down to prevent them from interfering with signals conveyed by subsequent releases of neurotransmitters. Some just leak away from the synapse into the fluid that surrounds nerve cells, while others are actively broken down by enzymes (**right, above**), and still others are reabsorbed for reuse (**far right**).*

*In order to **minimize the risk of running out**, neurotransmitters have to be made constantly. One site of manufacture is in the cell body where they are assembled under control of instructions from genes held in the DNA in the nucleus. Instructions are first copied onto an RNA molecule (**below**), which leaves the nucleus for an assembly site – a ribosome – on the endoplasmic reticulum. Another site of manufacture is in the axon terminal, where sub-units of neurotransmitters are put together.*

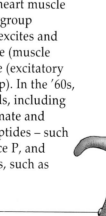

DNA

RNA

Dendrite

Like water from a mop and bucket, some neurotransmitters are released to do their work, then gathered up again for reuse.

NEUROTRANSMITTERS: DECADES OF DISCOVERY

Since neurotransmitters were first discovered, each decade has seen the investigation of a different group. First came acetylcholine, a major one in peripheral nerves. It excites at synapses where motor neurons connect to muscles, making them contract, but inhibits at heart muscle synapses, reducing the heart's activity. In the 1950s, amine-group neurotransmitters were found, including norepinephrine (excites and inhibits muscle and internal-organ function), dopamine (muscle movements, arousal, excitatory in the brain), histamine (excitatory in the brain), and serotonin (mood, appetite, pain, sleep). In the '60s, some were found to be amino acids or similar chemicals, including GABA and glycine (inhibitory in the brain), and glutamate and aspartate (excitatory in the brain). In the '70s, neuropeptides – such as endorphins and enkephalins, somatostatin, substance P, and CCK – were investigated. In the '80s came simple gases, such as nitrous oxide, and the energy molecule ATP.

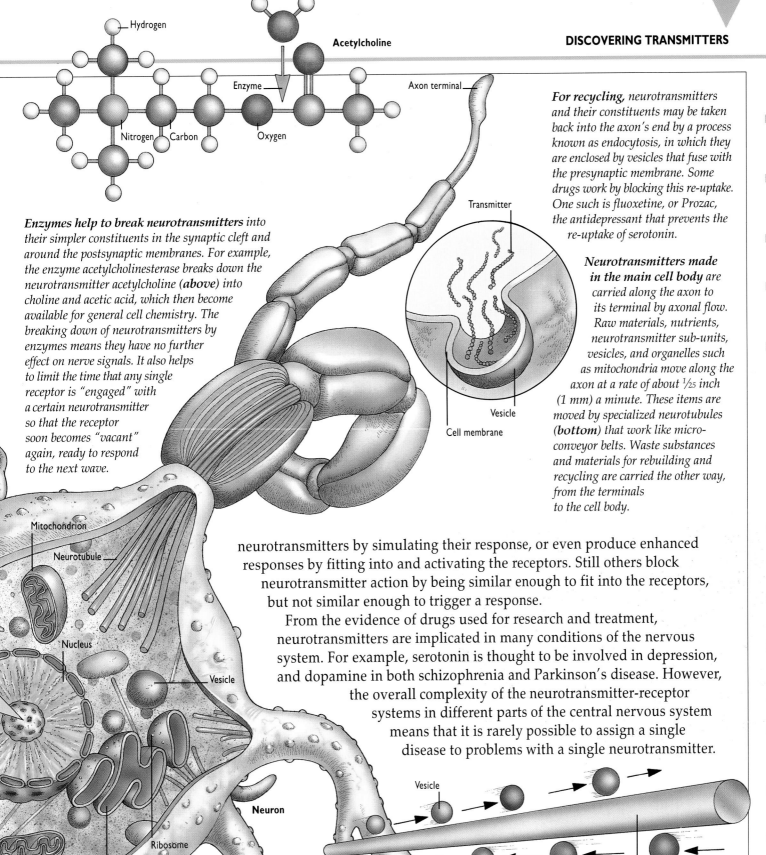

Hydrogen

Acetylcholine

Enzyme

Nitrogen Carbon

Oxygen

Axon terminal

Transmitter

Vesicle

Cell membrane

Mitochondrion

Neurotubule

Nucleus

Vesicle

Neuron

Ribosome

Endoplasmic reticulum

Vesicle

Neurotubule

Enzymes help to break neurotransmitters into their simpler constituents in the synaptic cleft and around the postsynaptic membranes. For example, the enzyme acetylcholinesterase breaks down the neurotransmitter acetylcholine (*above*) into choline and acetic acid, which then become available for general cell chemistry. The breaking down of neurotransmitters by enzymes means they have no further effect on nerve signals. It also helps to limit the time that any single receptor is "engaged" with a certain neurotransmitter so that the receptor soon becomes "vacant" again, ready to respond to the next wave.

For recycling, neurotransmitters and their constituents may be taken back into the axon's end by a process known as endocytosis, in which they are enclosed by vesicles that fuse with the presynaptic membrane. Some drugs work by blocking this re-uptake. One such is fluoxetine, or Prozac, the antidepressant that prevents the re-uptake of serotonin.

Neurotransmitters made in the main cell body are carried along the axon to its terminal by axonal flow. Raw materials, nutrients, neurotransmitter sub-units, vesicles, and organelles such as mitochondria move along the axon at a rate of about $1/25$ inch (1 mm) a minute. These items are moved by specialized neurotubules (*bottom*) that work like micro-conveyor belts. Waste substances and materials for rebuilding and recycling are carried the other way, from the terminals to the cell body.

neurotransmitters by simulating their response, or even produce enhanced responses by fitting into and activating the receptors. Still others block neurotransmitter action by being similar enough to fit into the receptors, but not similar enough to trigger a response.

From the evidence of drugs used for research and treatment, neurotransmitters are implicated in many conditions of the nervous system. For example, serotonin is thought to be involved in depression, and dopamine in both schizophrenia and Parkinson's disease. However, the overall complexity of the neurotransmitter-receptor systems in different parts of the central nervous system means that it is rarely possible to assign a single disease to problems with a single neurotransmitter.

See also

BUILDING THE BRAIN
▶ To fire or not to fire? 48/49

▶ Crossing the gap 52/53

▶ On or off 56/57

▶ Unlocking the gate 58/59

▶ Remaking the mind 62/63

SURVEYING THE MIND
▶ Brain maps 30/37

INPUTS AND OUTPUTS
▶ Feeling pain 104/105

▶ Rhythms of the mind 120/121

STATES OF MIND
▶ The resting mind? 160/161

▶ Physical cures 166/167

▶ Feeling low 168/169

▶ Anxious states 172/173

Remaking the mind

The brain changes itself throughout life, as it thinks, controls the body, learns, and makes memories.

A living, changing, evolving entity, the brain adapts itself continuously according to its inputs, internal processing, and outputs. These changes occur within and between neurons, in various ways, and over various timescales. Neurons can alter the strength of their connections and thus the way they respond to or send messages. These messages are transmitted in the form of neurotransmitter molecules released into the synapses – the gaps where signals are passed between neurons.

The change in strength can be short-term when, for instance, the amount of neurotransmitter released into the synapse by the sending cell rises during a burst of signaling. A change can also be longer term when a given amount of neurotransmitter arriving at the receiving cell has a greater effect than originally. Long-term changes occur over periods ranging from a few hours to weeks and can involve changes between the connecting structures of neurons.

Grains of sand are rearranged in their billions as they are blown by the wind, endlessly reshaping the dunes and altering the desertscape. Similarly, the billions of neurons in the brain continually make and remake their connections on many levels, in response to their individual histories and current activities, to reshape the landscape of the mind.

THE SHRINKING SEA SNAIL

You shrink away if something possibly harmful touches you. So does *Aplysia*, a type of marine snail. Touch it lightly on its siphon (breathing tube) and its sensory neurons pick up the signal; the message passes via axons and synapses to neurons controlling the gill muscles, and the gill is withdrawn for safety. But after a few touches, the snail habituates, no longer reacting to the touch.

If an electric shock is then applied to its tail, the snail is sensitized for some time, and even a gentle touch to the siphon brings about an increased withdrawal. The nerve signals traveling along axons from the tail's sensory neurons pass via intermediate interneurons to facilitate (increase the effect of) signals coming from the siphon's neurons. This change in the way a set of neurons works is called presynaptic facilitation, since the effect takes place before the signal reaches the synapse that conveys the withdrawal message.

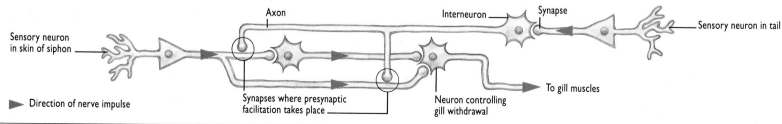

Sensory neuron in skin of siphon

Axon

Interneuron

Synapse

Sensory neuron in tail

To gill muscles

▶ Direction of nerve impulse

Synapses where presynaptic facilitation takes place

Neuron controlling gill withdrawal

Some synapses can change the strength of their connection. When a regular number of firings, or action potentials, travel along a neuron's axon, they induce a corresponding postsynaptic potential in the receiving neuron's membrane. But when a very rapid burst of firings arrives, as in a burst of nerve signals, the resultant postsynaptic membrane potential changes overall, rising in strength, due to an increase in neurotransmitter release. For a while, perhaps up to a few minutes or even hours, the postsynaptic potential – and thus the likelihood of the receiving neuron itself firing – is increased even after a return to a regular rate of firing. The synapse has been potentiated.

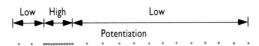

Presynaptic firing rate

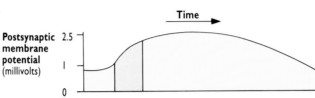

Time

Postsynaptic membrane potential (millivolts)

Postsynaptic potentials in the membrane of a receiving neuron are low when the presynaptic firing rate is low. If the rate rises, the resulting postsynaptic potentials can rise. This effect may last so that after the rate has dropped, each firing still gives a higher postsynaptic potential.

See also

BUILDING THE BRAIN
▶ Crossing the gap 52/53

▶ On or off 56/57

▶ Unlocking the gate 58/59

▶ Discovering transmitters 60/61

▶ Recovering from damage 64/65

▶ Linking up 74/75

INPUTS AND OUTPUTS
▶ On the scent 108/109

FAR HORIZONS
▶ Conditioning the mind 128/129

▶ The infinite store? 130/131

▶ Active memory 132/133

STATES OF MIND
▶ Physical cures 166/167

The most common excitatory neurotransmitter in the brain is glutamate, a chemical that activates several subtypes of receptors. These occur in large numbers in the hippocampus, part of the brain involved in memory and learning. Among them are NMDA (N-methyl-D-aspartate) receptors, which, like other such receptors, need both glutamate and an agonist, or "helper," chemical to generate local postsynaptic potentials – in this case glycine. In a presynaptic neuron (**below left**) that is quiescent, some activity takes place, but the neuron fires action potentials only occasionally, like a clock ticking in the background. These generate corresponding but low-level depolarizations in the postsynaptic membrane. The main depolarization changes are due to the inflow of sodium ions (Na^+). The NMDA receptor is non-operational under these conditions.

When, like a clock setting off its alarm, the presynaptic neuron sends a rapid burst of signals to the synapse (**right**),

a build-up of local depolarization results. The NMDA receptor, being voltage-dependent, responds. A magnesium ion (Mg^{2+}) "plug" which blocks its channel is driven out. This allows both sodium ions and calcium ions (Ca^{2+}) to flow through, giving a stronger depolarization in the receiving cell. This effect can last some time. Inside the postsynaptic neuron, calcium may also be involved in microstructural changes that underlie long-term potentiation.

This is just one of the vast array of processes whereby neurons alter and adapt their activities, changes which underlie thinking, learning, and other mental processes in the brain.

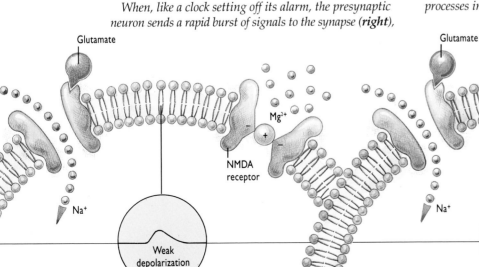

Recovering from damage

Once an adult loses a brain cell, it is gone forever. But unless the injury is severe, the brain can restore at least some function.

Most organs, such as the skin or liver, can usually repair damage because the cells in their tissues are all similar and can proliferate and replace those lost. But the fully developed brain presents problems, because every neuron is unique and so specialized in form and function that the cells can no longer proliferate. With its snakelike axon and thousands of dendrites synapsing with the other neurons, its individual three-dimensional architecture is impossible to replace. So repairing brain damage, when possible, is a slow and difficult process.

Since neurons in the brain cannot be replaced, people with brain damage who undergo rehabilitation have to try to "retrain" undamaged neural circuits to take on new roles, such as controlling the muscle movements of walking. This involves the growth of axons and dendrites on existing neurons, to make new connections as they rewire themselves. Such activity is the basis of the developing brain. But in a mature brain, it is a more modest process. There are, for instance, the supporting glial cells, or astrocytes, which form fibrous tissue at the damage site, creating an impenetrable microscopic thicket for any axons and dendrites trying to grow. Yet there can be progress. Vacant synapses are thought to produce a substance that stimulates nearby axons to sprout new branches and grow toward them. Growth usually stops after a few micrometers because of the hostile conditions, so larger-scale repair may have to take place in a series of short steps as the brain commandeers some of its neural circuits for new uses.

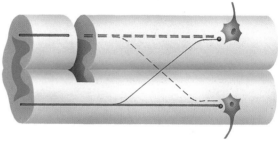

In the spinal cord there are many axons crossing from one side to the other that appear to have no function. However, if one side of the cord is damaged, they come into their own and can restore contact to a neuron below the site of the damage, and some function can be regained.

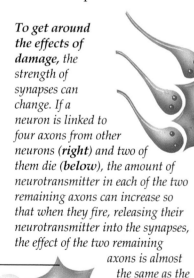

To get around the effects of damage, the strength of synapses can change. If a neuron is linked to four axons from other neurons (**right**) and two of them die (**below**), the amount of neurotransmitter in each of the two remaining axons can increase so that when they fire, releasing their neurotransmitter into the synapses, the effect of the two remaining axons is almost the same as the original four.

Neuron

Defunct axon

Neurotransmitter

Synapse

Axon

The receiving part of a neuron (above) can become more sensitive after the loss of an axon. The number of receptors can increase in response to a lower stimulation from just a single axon (**below left**). With more receptors, the receiving neuron is able to make up for the effect of reduced neurotransmitter release.

Receptor

A neuron might be connected to two axons, each splitting into two (**below left**). If one of the axons is destroyed, another axon can, in some circumstances, grow new terminals as replacements (**below right**).

When an axon has been severed (dotted line), it can put out a so-called regenerative sprout to link up with another neuron close by and thus restore some function. It will never be quite as good as new, however.

An earthquake shatters the landscape and breaks lines of communication and transportation, especially roads. The blockage can be passed by diverting along rough tracks through less damaged areas. In the long term, it may be worth developing these temporary tracks into permanent roads, rather than repairing the original routes.

When its neurons are damaged, the nervous system also establishes diversions and bypasses, sending nerve signals along them – either until the original damage is repaired or on a permanent basis.

The process of overcoming damage in the brain can be slow in adults, and full recovery is unlikely. The same is not true of children, however. This is because the brain is still developing in children – its neurons are all established before birth, but the axons and dendrites continue to grow, forming new connections all the time. So a young child's brain can cope with relatively severe damage, not only rerouting its links to avoid that damage, but actually regrowing some of the connections. The adult brain is unable to do this because it is already fully formed and cannot continue its growth.

A CASE OF DAMAGE

A man was accidentally hit hard on the head with a baseball ball and was unconscious for a short while but seemed to recover. He was taken to hospital where an X-ray showed he had a fracture. A few hours later, he developed a headache, felt nauseous, vomited, and became drowsy. He was quickly given a computer tomography (CT) scan which revealed an epidural hemorrhage.

In an epidural hemorrhage, the blood vessels in the dura mater, the outermost of the three layers (meninges) covering the brain, rupture. When these large vessels bleed, blood rapidly fills the space between the brain and skull (as shown in the red oval feature at center left of the CT scan below). This is a dangerous complication that quickly puts the brain under pressure and which can starve it of oxygen from its own blood supply. The patient recovered after an operation to relieve the pressure and repair the blood vessels.

Apart from such accidents, the most common condition in which the brain is starved of blood is stroke, when the blood supply to part of the brain is cut off or blood leaks from its vessels into the brain tissue. The symptoms include slurred speech, loss of muscle control, numbness, paralysis, confusion, and disruption of the senses such as touch or sight. Among the more common causes are cerebral thrombosis, a blood clot that forms in an artery supplying the brain; cerebral embolism, a blood clot, fatty lump, or other object that lodges in an artery supplying the brain; aneurysm, a weak spot in the arterial wall that can burst; and damage caused either by disease or by an injury.

If the stroke affects only one side of the brain, then only the other side of the body is affected. With time and physiotherapy, the stroke victim who survives the initial trauma can recover much lost function as the brain finds alternative pathways around the damaged regions.

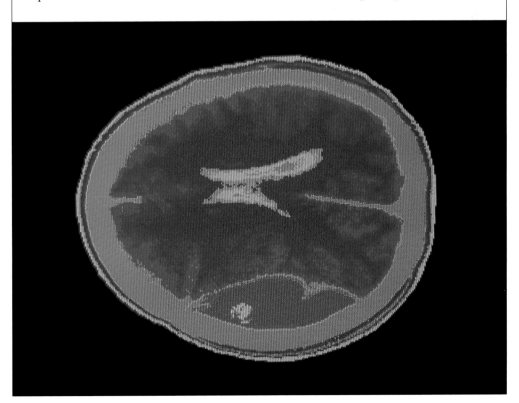

See also

BUILDING THE BRAIN
▶ Support and protection
42/43

▶ The brain's cells
44/45

▶ Remaking the mind
62/63

▶ The cell factory
72/73

▶ Linking up
74/75

▶ The aging brain
76/77

SURVEYING THE MIND
▶ Discovering through damage
24/25

▶ The view from outside
28/29

▶ Brain maps
30/37

INPUTS AND OUTPUTS
▶ Body links
80/81

▶ The long junction
82/83

STATES OF MIND
▶ The failing mind
176/177

Food for thought

Like the body's other organs, the brain depends on two vital substances, brought to it by circulating blood.

Along with the glucose, or blood sugar, from digested food which powers life processes, we also need oxygen from the lungs. It is an essential ingredient in the chemical pathways of cellular respiration, the process of extracting energy from glucose in a form that cells can use. The brain is by far the body's "hungriest" part – the major consumer of glucose and oxygen. Pound for pound, it requires 8 to 10 times more glucose and oxygen than other organs. Indeed it uses almost 20 percent of the body's oxygen intake, despite being only 2 percent of total body weight. But why?

Unlike an obviously contracting muscle or our writhing intestines, the brain does not look outwardly active. Its activities are at the electrochemical and molecular levels, as its billions of neurons constantly maintain resting potentials, alter them to fire nerve signals, and synthesize neurotransmitters. The various ion pumps and other cellular machinery involved in these processes use large amounts of energy – and they need it continually, whether the body is running a sprint or fast asleep. They receive it at a constant rate – blood flow to the brain along the cerebral arteries is 1½ pints (750 ml) each minute. An equivalent amount of electrical energy would run a 10-watt light bulb – perhaps not very bright, but in bodily terms a huge amount. And to maintain this steady supply of energy, if glucose levels in the blood drop, other organs reduce their consumption of it, leaving more fuel for the brain.

Blood flow to organs fluctuates greatly according to their needs and general body activity: when we are resting, they require less blood; during physical activity, they need more. But the brain's need for blood is different. It remains almost constant, flowing through at 1½ pints (750 ml) per minute no matter what the brain or body is doing – from resting to intense thought, such as reading a book, or hard physical exercise.

The brain needs this constant supply because unlike the liver, which can store energy as glucose and starch, and the muscles, which can store oxygen as myoglobin, it cannot store its fuel. And just 5 to 10 seconds without oxygen flowing to the brain causes a person to lose consciousness. A further 10 seconds and body muscles begin to twitch convulsively as the brain's motor centers lose control of them. After four minutes, serious damage begins in the neurons and other brain cells. Just 10 minutes after the initial loss of supply, this damage is usually irreversible – and fatal.

Like a leaky boat in which the crew constantly bails out water to keep it afloat, the brain uses most of its energy – about 80 percent – to pump sodium ions out of nerve cells and potassium ions in, against their natural concentration levels. The pump uses the energy molecule ATP, which turns to ADP and phosphate (P_i) in the process.

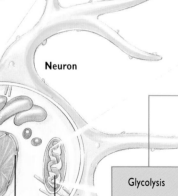

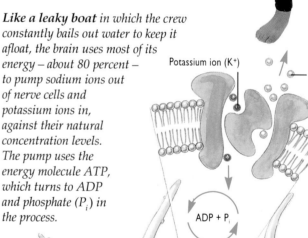

Potassium ion (K⁺)

Sodium ion (Na⁺)

Detail of sodium pump

Cell membrane

ADP + P_i

Neuron

The brain is protected *from natural body wastes, drugs, and other toxins which might seep through the walls of its capillaries by a special barrier – the blood–brain barrier. This is formed by tight, secure junctions between the endothelial cells in its capillary walls. But the barrier might also keep out useful nutrients and chemicals. So special proteins in the cell membranes of the endothelial cells recognize such nutrients and transport them from the capillary out to neurons and other brain cells.*

Endothelial cell

Nucleus

Capillary in brain

Mitochondrion

Vesicle

Glycolysis

2
ATP

Red blood cell

Tight junction

Astrocyte end foot

Dendrite

Nucleus

Mitochondrion

Axon

See also

**BUILDING
THE BRAIN**
▶ Building blocks
40/41

▶ Support and
protection
42/43

▶ The electric cell
46/47

▶ Recovering
from damage
64/65

▶ The aging
brain
76/77

**SURVEYING
THE MIND**
▶ Discovering
through damage
24/25

**INPUTS AND
OUTPUTS**
▶ Keeping control
116/117

▶ Dealing with
drives
118/119

**STATES OF
MIND**
▶ Abnormal
states
162/163

WHEN THE FUEL SYSTEM BREAKS DOWN

Generally perfectly healthy and sane, a man found that he was experiencing extremely peculiar thoughts and moods. Things came to a head when he became abnormally fascinated with a drill which he was using to put up some shelves, to the extent that his terrified wife believed he was about to harm himself and telephoned for a doctor. It turned out that he was suffering from a condition known as hypoglycemia, in which glucose in the blood drops to a dangerously low level, thus effectively starving the brain of fuel.

If the brain is starved in this way, loss of consciousness quickly results – just as is the case with a failure in the oxygen supply. What this man was experiencing was a state that occurs between consciousness and unconsciousness, in which sufferers commonly report aggressive feelings and loss of self-control, free will, and the ability to follow any sort of reasoning – symptoms not unlike those of schizophrenia.

There is a simple cure, however. By monitoring levels of glucose in the blood, sufferers can tell if they are falling too low and thus prevent an attack by eating something that will raise them again. Sufferers need to be aware of such things as when they last had a meal, how much they ate, and whether they exercised after eating.

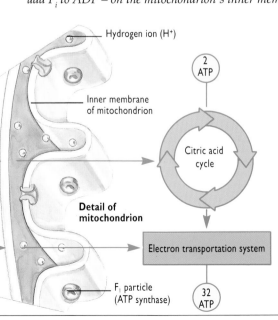

In all body cells, including neurons, glucose – a sugar with energy locked up in its chemical bonds – is broken apart and the energy released in a series of biochemical stages. At each stage, energy is "captured" by adding a phosphate group (P_i) to the substance ADP (adenosine diphosphate) to make ATP (adenosine triphosphate), a high-energy molecule that powers many cell processes.

In the glycolysis stage, there is a net gain of two ATPs. In the citric acid (Krebs) cycle, which takes place inside the mitochondria, two more ATPs are gained. In the electron transport system stage, powered by energy from hydrogen ions (H^+) released in glycolysis and the citric acid cycle, the gain is 32 ATPs. These activate F1 particles – enzymes that add P_i to ADP – on the mitochondrion's inner membrane.

Hydrogen ion (H^+)

Inner membrane
of mitochondrion

**Detail of
mitochondrion**

F_1 particle
(ATP synthase)

2
ATP

Citric acid
cycle

Electron transportation system

32
ATP

***Different organs use blood at
varying rates*** depending on the
level of activity in the body, as these
figures show (all are per minute).

Blood used by brain
*At rest 1½ pints (750 ml)
Exercising 1½ pints (750 ml)
Digesting 1½ pints (750 ml)*

Blood used by muscles
*At rest 2½ pints (1.2 liters)
Exercising 3¼ gallons
(12.5 liters)
Digesting 2 pints (1 liter)*

Blood output/used by heart
*At rest 12 pints/8½ ounces
(5.8 liters/250 ml)
Exercising 4½ gallons/
8½ ounces (17.5 liters/250 ml)
Digesting
2 gallons/10 ounces
(7.3 liters/300 ml)*

***Blood used by
intestines***
*At rest 3 pints
(1.4 liters)
Exercising
20 ounces
(600 ml)
Digesting
7½ pints
(3.5 liters)*

The brain plan

Along with the rest of the body, the brain grows and matures under the control of the genetic blueprint.

If you could look back at yourself when you were a day-old fertilized egg, you would see a tiny dot of living tissue, only just visible to the unaided eye. You would be a single cell about $1/250$ inch (0.1 mm) across, formed when an egg cell from the mother joins with a sperm cell from the father. Fertilization marks the time when genes from both parents form the complete genetic make-up of a new individual.

Genes provide the instructions for the development and functioning of a human body in its entire and intricate detail. There are between 100,000 and 200,000 of them in the full gene set, the human genome. Each individual gene takes the form of a functional section, or sub-unit, of one of the 23 pairs of DNA molecules contained in the nucleus of every body cell.

After a short time, the fertilized egg divides into two cells, then four, and so on. Amazingly, each time a cell divides, the entire set of DNA molecules is copied, so that each of the new daughter cells receives the full gene set. At first the cells all look the same but, gradually, as the number increases, they differentiate. Differentiation is achieved when only certain genes in a cell are activated, such as the genes needed to make a nerve cell, and the rest of the set remains quiescent. Groups of cells take on their own structure and function, as muscle cells, skin cells, nerve cells (neurons), and so on. In the embryo the process of cell differentiation determines the order and position in which structures appear. It takes place "top-down," with the brain leading the way, followed by the head, face, and special sense organs, then the heart, intestines, and other internal organs, the arms and, finally, the legs.

As a fertilized egg drifts along the Fallopian tube to the uterus, it divides every 15 to 20 hours. At four days, it is known as a late blastocyst, a ball of 64 identical cells. At one week, it has implanted in the uterus. The blastocyst's cells invade the uterine tissues – richly supplied with blood by capillaries – to obtain nutrients for growth.

We all start out with the same potential, in theory, but we do not stay the same. Thus one person may end up an architect and another a manual worker. Similarly, every cell in an embryo has a full set of genes and so the same theoretical potential, yet the end result is a wide range of specialized cell types.

At about two weeks the body of the embryo is a two-layered plate of cells, with the yolk sac on one side and the protective fluid-filled amniotic cavity on the other. A week later, the embryonic disk is evident.

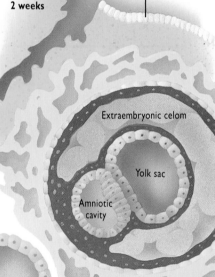

2 weeks

Uterine wall

Extraembryonic celom

Yolk sac

Amniotic cavity

Uterine glands

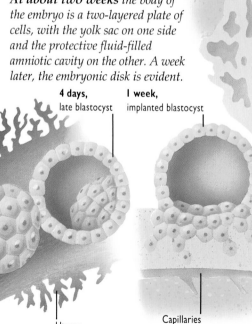

4 days, late blastocyst

I week, implanted blastocyst

3 days, morula

2½ days, 8 cells

2 days, 4 cells

I day, 2 cells

Fallopian tube

Uterus

Capillaries

See also

BUILDING THE BRAIN
▶ The developing brain 70/71

▶ The cell factory 72/73

▶ Linking up 74/75

SURVEYING THE MIND
▶ Brain maps 30/37

INPUTS AND OUTPUTS
▶ Body links 80/81

▶ The long junction 82/83

▶ Keeping control 116/117

FAR HORIZONS
▶ First thoughts 142/143

STATES OF MIND
▶ The resting mind? 160/161

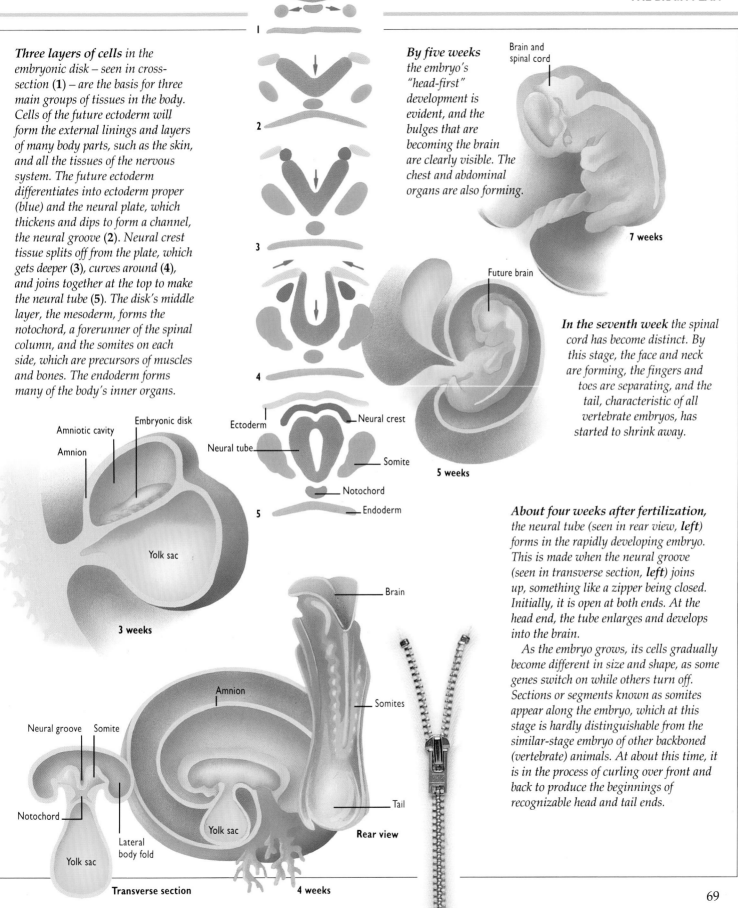

Three layers of cells in the embryonic disk – seen in cross-section (**1**) – are the basis for three main groups of tissues in the body. Cells of the future ectoderm will form the external linings and layers of many body parts, such as the skin, and all the tissues of the nervous system. The future ectoderm differentiates into ectoderm proper (blue) and the neural plate, which thickens and dips to form a channel, the neural groove (**2**). Neural crest tissue splits off from the plate, which gets deeper (**3**), curves around (**4**), and joins together at the top to make the neural tube (**5**). The disk's middle layer, the mesoderm, forms the notochord, a forerunner of the spinal column, and the somites on each side, which are precursors of muscles and bones. The endoderm forms many of the body's inner organs.

By five weeks the embryo's "head-first" development is evident, and the bulges that are becoming the brain are clearly visible. The chest and abdominal organs are also forming.

In the seventh week the spinal cord has become distinct. By this stage, the face and neck are forming, the fingers and toes are separating, and the tail, characteristic of all vertebrate embryos, has started to shrink away.

About four weeks after fertilization, the neural tube (seen in rear view, **left**) forms in the rapidly developing embryo. This is made when the neural groove (seen in transverse section, **left**) joins up, something like a zipper being closed. Initially, it is open at both ends. At the head end, the tube enlarges and develops into the brain.

As the embryo grows, its cells gradually become different in size and shape, as some genes switch on while others turn off. Sections or segments known as somites appear along the embryo, which at this stage is hardly distinguishable from the similar-stage embryo of other backboned (vertebrate) animals. At about this time, it is in the process of curling over front and back to produce the beginnings of recognizable head and tail ends.

Labels:
Brain and spinal cord

7 weeks

Future brain

5 weeks

Amniotic cavity
Amnion
Embryonic disk
Yolk sac
3 weeks

Ectoderm
Neural crest
Neural tube
Somite
Notochord
Endoderm

Brain
Amnion
Somites
Tail
Rear view

Neural groove
Somite
Notochord
Lateral body fold
Yolk sac
Yolk sac
Transverse section
4 weeks

The developing brain

From its tiny embryonic beginnings, the brain grows rapidly, reaching a quarter of its adult size by birth.

About eight weeks after egg joined sperm to create a genetically unique individual, the pinpoint-sized single cell that was the fertilized egg has divided, its cells have differentiated, the early embryo has passed through various tadpolelike stages, and a miniature, recognizably human body has developed. At this time, it is about as big as an adult's thumb, yet all the main body parts and organs have appeared, right down to fingers and toes. The head is grape-sized and still slightly larger than the whole of the torso, since the embryo follows a head-first order of development. The brain occupies most of this head volume, with the facial features tucked below it, still small and lacking in detail.

The end of the eighth week marks a sudden change, but in name only. By scientific convention, the embryo is now referred to as a fetus. It continues to grow at an incredible rate, fuelled by oxygen and nutrients from the mother's blood obtained via the placenta and umbilical cord. While it seems to us that a newborn baby grows fast in its first few months, it is far outstripped by the fetus. At its maximum rate of size increase, the fetus is enlarging 10 times more quickly than a newborn baby. The fetal brain grows even faster. The lower portions, such as the medulla, are among the first to form. These are the more basic parts and deal with mechanical body functions and life-support systems such as heartbeat. The areas such as the cerebrum initially lag behind.

At four months, the brain's full general structure is present. By the fifth month, the fetus can grip, kick, and hiccup, showing that reflexes and muscle control (motor) nerves are functioning. At seven months, the fetus practices controlled, rather than reflex, movements such as swallowing, pushing,

At birth a baby mammal leaves its mother's womb through the birth canal, passing through the gap in the pelvic girdle, the bowl-shaped "hipbone" at the base of the abdomen. In most mammals, the head is relatively small and slips out easily; baby dolphins are usually born tail-first. Since dolphins are supported by water, they have developed a relatively wide birth canal, so the baby dolphin's brain can be almost the full adult size at birth. In humans, since we are two-legged walkers, the mother's anatomy has evolved to put a limit on the birth canal opening. So head size, and thus brain size, is restricted.

and "breathing" as it floats in its protective pool of amniotic fluid. From this time, too, the upper brain begins to enlarge fast, and the cerebral hemispheres continue to develop their typically wrinkled appearance. At nine months, the fetus is full term, ready to be born. It is an amazing change from the original single cell, the fertilized egg. This anatomical development is the result of astonishing microscopic flurries of cell multiplication and migration as a new human being – equipped with a functioning brain – is prepared for the outside world.

At three months, the spinal cord of the fetus has reached its definitive structure, with an H shape of gray matter running along inside the white matter. The cerebellum and cerebrum (part of the forebrain) are still fairly smooth, with few of the wrinkles characteristic of the adult brain. The fetus is about 2 inches (5 cm) long, measured from crown to rump, that is, top of head to bottom, and is lying in the typically curled "fetal position."

At seven weeks, the embryo is about ⅔ inch (1.7 cm) long. The general areas of the brain – hindbrain, cerebellum, mid-, and forebrain – are clearly formed. The cerebral hemispheres (in the forebrain), which will make up 90 percent of the brain's total adult volume, now constitute only about 10–15 percent of the total.

Midbrain

Cerebellum

Forebrain

Hindbrain

Spinal cord

Eye

7 weeks

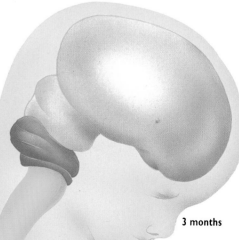

3 months

COLONIZING CELLS

When the neural tube is forming early in an embryo, two strips of cells are "pinched off" from the edges of the neural folds and finally flank the tube. While the neural tube goes on to form the central nervous system (CNS) – brain and spinal cord – these strips of cells, which will form part of the autonomic nervous system (ANS), multiply and migrate in waves. Known as neural crest cells, they make their way between other cell groups to set up colonies in other areas. In addition to forming parts of the ganglia and nerves of the ANS, they also make up pigment cells in skin, enamel-making cells in teeth, and connective tissue bodywide.

Neural tube

Spinal ganglion

Spinal ganglion

Spinal nerve

Spinal cord

Abdominal aorta

Migrating autonomic nervous system cells

Cross section of embryo

Migrating neural crest cells leave behind colonies that develop into various body tissues, both nervous and other tissues. They include sensory neurons in the nerves connecting to the spinal cord, and neurons of the ANS that run from their ganglia to inner organs such as the heart, blood vessels, and intestines.

See also
BUILDING THE BRAIN
▶ The brain plan 68/69

▶ The cell factory 72/73

▶ Linking up 74/75

SURVEYING THE MIND
▶ Comparing brains 18/19

INPUTS AND OUTPUTS
▶ The long junction 82/83

▶ Keeping control 116/117

FAR HORIZONS
▶ First thoughts 142/143

▶ Male and female 148/149

STATES OF MIND
▶ The resting mind? 160/161

At six months, growth and development of the fetus's body has speeded up and is fast catching up with the head region. The total length is 14 inches (35 cm). In the brain, the cerebral cortex is beginning to dominate as the fastest-growing region, developing its typical convolutions on the surface.

At nine months, the baby's brain looks much like the adult version. It is still more developed than other body parts, such as muscle and bone, as well as being disproportionately large. The midbrain has been obscured by the relatively huge cerebral cortex which forms most of the forebrain.

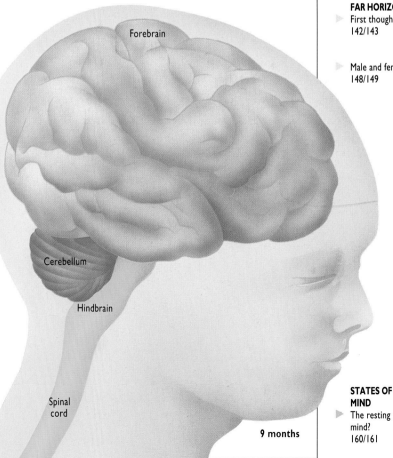

Forebrain

Cerebellum

Hindbrain

Spinal cord

6 months

9 months

The cell factory

Before birth, many billions of nerve cells are created that have to find the way to their correct place in the brain.

In a newborn baby, the neurons are more or less complete in number and in their basic positions within the brain. But they are not yet mature, in that their connections with other neurons are not fully established. They have all originated from cells of the layer lining the neural tube, which is present in a three-week-old embryo and is the precursor of the brain and spinal cord. They are produced by the standard cell-division process – mitosis – which gives rise to "pre-neurons," or neuroblasts. During an embryo's early weeks and months, the rate of cell multiplication is enormous – 200,000 or more neuroblasts are formed every second. At any one time, millions of them are migrating to their correct positions, controlled by a massively complex system of guiding factors. One of these involves "sticky" regions on cells known as cell adhesion molecules (CAMs). If two cells have similar CAMs on their surfaces, they are more likely to stay together. So one set of cells can form the route or base for another group.

There are also the so-called cell-substrate adhesion processes, in which neuroblasts attach to and travel along non-cellular lattices, scaffoldings, and membranes of proteins, sugars, and other molecules, formed in the extracellular environment. Migrating nerve cells are also affected by soluble chemicals such as hormones, growth factors, and neurotransmitters being manufactured by their neighbors.

As each group of neuroblasts settles into position, some of their genes switch off, while others switch on. The cells begin to change in shape and function and start to mature into neurons, putting out their dendrites and axons. The growth of these signifies the next phase in brain development: neurons linking up with one another.

A tree trunk's growth rings indicate its age in years, with two rings for each year. The light one reflects fast spring and early summer growth, the dark one late summer—fall slowdown. Each annular ring is the result of a layer of cells, the cambium, specialized to divide and make more cells. It produces thick heartwood cells on its inside, for water transport, and softer sapwood cells on its outside, just under the bark, for nutrient and sap transport. As heartwood accumulates yearly, the trunk thickens. In a similar way, layers of cells in the brain multiply and relocate to take up set positions.

The cerebral neocortex is the site of our most sophisticated mental processes, including thinking and awareness. In the five-week-old embryo, it is just the microthin wall of pocket-like outgrowths from the sides of the forebrain. It develops three layers. In the neuroepithelial layer, the cells divide, producing neuroblasts which find their way to their brain positions by moving up guiding radial glioblasts.

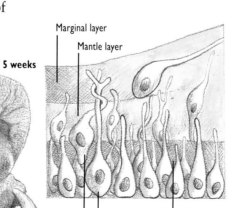

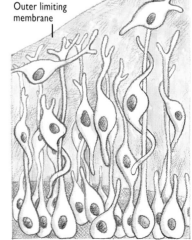

5 weeks

3 months

Outer limiting membrane

Marginal layer

Mantle layer

Radial glioblast

Neuroblast

Neuroepithelial layer

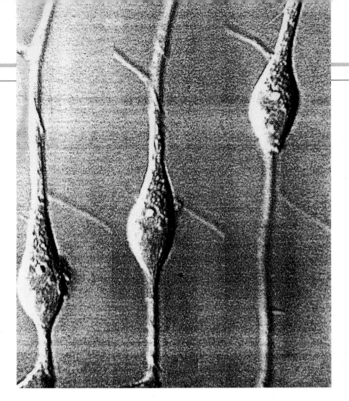

The neuroblasts that will mature into "gray matter" neurons of the cerebral cortex find their way partly with the aid of glial cells. Glia generally provide a cellular system of scaffolding, support, protection, and nourishment for the brain. Radial glioblast cells, along the inner neuroepithelial layer lining the neural tube, grow long, thin extensions outward, away from the center of the tube, like the spokes of a wheel. Each neuroblast oozes and crawls along this "glial monorail" **(left)** and alights at its predetermined resting position.

The process happens in waves to form the six cortical layers. The innermost layer (VI), which is nearest to the neuroepithelial layer, forms first. This means that successive waves of neuroblasts must crawl between their already settled counterparts, in order to establish their own layer on the outer side. It is like concentric rings of suburbs growing ever outward around a city center, using trains and roads as their transportation system.

See also

BUILDING THE BRAIN
▶ Building blocks 40/41

▶ Support and protection 42/43

▶ The brain's cells 44/45

▶ The brain plan 68/69

▶ The developing brain 70/71

▶ Linking up 74/75

FAR HORIZONS
▶ First thoughts 142/143

STATES OF MIND
▶ Individual minds 184/185

In a five-month embryo the cortex is developing its characteristic six layers. Successive waves of neuroblasts migrate from the innermost layer to form the outermost.

By eight months the basic physical structure of the brain, including the neocortex, is almost complete. To form each of the six cortical layers, a migratory wave of neuroblasts has moved out along the pathfinding extensions of the radial glioblasts and settled in place before differentiating into the more mature neurons characteristic of that particular layer: fusiform in VI, pyramidal in V, stellate in IV, and so on. The outermost layer (I) forms last. The axons of all the layers gather to form the white matter below the cortex. Some glial cells of the original neuroepithelial layer now form the ependymal layer, the barrier lining the brain's fluid-filled ventricles, as shown in the cross section **(below)** of the eight-month-old fetal brain.

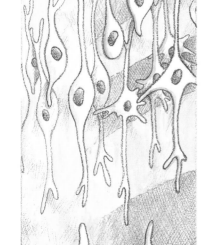

5 months

8 months

Glial cell

Neocortical layers

I

II

III

IV

V

VI

Stellate neuron

Pyramidal neuron

Fusiform neuron

Ependymal cell

Subcortical white matter

Neocortical layers

Ependymal layer

Ventricle

Subcortical white matter

73

Linking up

Without connections, neurons could not function. How they link is one of the growing brain's most extraordinary stories.

In its first year of life, a baby's brain triples in size, to almost three-quarters of its adult dimensions. Growth then slows dramatically, with full adult bulk attained at about 17 years. This huge and rapid initial growth in brain volume is due chiefly to the insulating myelin sheaths forming around nerve fibers. But there is another factor: the growth of neurites – dendrites and axons – from the nerve cell bodies to form links, or synapses, with one another. The dendrites elongate and snake between other cells to near neighbors. Many axons undergo incredible growth, reaching past millions of other cells and tissues to seek their target cells, perhaps in a far-off organ.

Dendrites and axons find their way and link using a variety of cues and clues, from physical and positional to physiological, chemical, and electrical. And during these early years, the brain's connection network is "plastic" – it constantly re-wires, alters, and updates, according to the environment. But by the age of 16, its physiology is mature, and the molecular-cellular machinery for making new connections becomes much less efficient.

A RICHER PLACE TO BE

Researchers have discovered that rats raised in bare dark cages, with minimal sensory stimuli of any kind, develop a less sophisticated and less heavy brain cortex than normal. By contrast, those raised in a highly stimulating "enriched" environment, with plenty of extra sounds, sights, smells, tastes, and touches, have a more developed brain cortex than normal. This is thought to be because the brain is at its most adaptable and plastic stage just after birth, when animals – especially mammals – begin to explore and learn about their world. With stimulation, the brain cells make more connections with one another.

Humans respond in the same way and need the senses and mind to be stimulated, too. Babies in particular need such stimulation, and are thought to thrive in an environment which is full of bright objects and in which people talk to them – even if they are unable to understand the meaning of the words.

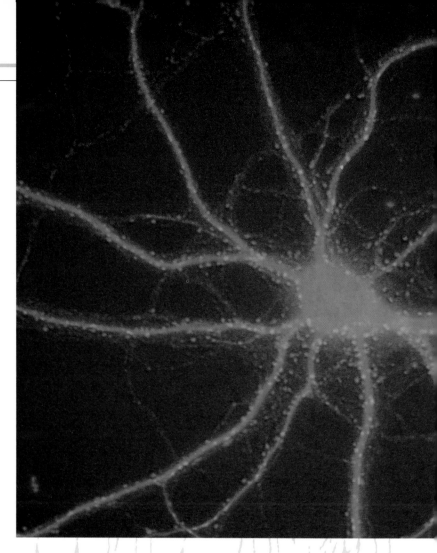

A nerve cell on a laboratory dish sends lengthening neurites into its surroundings, like an octopus growing new tentacles. Their growth can be encouraged and directed by various factors, such as trails of chemicals on the glass surface, dissolved chemicals seeping from one direction through the nutrient solution, and even physical features such as scratches or cracks in the glass.

During the first few years of a human child's life, its brain goes through a stage of incredible activity as numerous lines of communication are set up and the bodies of the nerve cells grow slightly. But the number of cells does not increase after birth. In these early years, the action is centered on creating connections – synapses – whose networks are shaped by current needs, imitation, memory, learning, and accumulating experience. Of course, this process then continues through life, although the pace is far slower in the mature brain.

Bees may not be bright, but they work hard. Much of their life and behavior is encoded in their genes. Worker bees follow strict instincts as they collect nectar from blooms and maintain the hive. In the same way, the earliest development of the human embryo follows the chemically coded instructions in the genes of its cells. These preprogram cells to multiply, move, and differentiate in certain ways, as the basic parts of the body and its nervous system take shape.

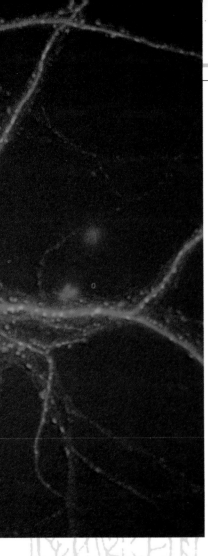

TURNING OFF THE SUICIDE SWITCH

A number of factors make a neuron's axons and dendrites, or neurites, link up with one another: genetic programming, pioneer neurons, guiding and signpost chemicals sent out by target neurons to "attract" axons, and a substance known as nerve growth factor (NGF). This is a soluble protein that encourages neurons to put out neurites in its direction, like ants following a scent trail, in search of the source of NGF.

Making such connections is central to the life of each cell, since cells have suicide switches which must be switched off if the cells are to survive. A cell that has no link to any others – and thus receives no chemical message to stay alive – will "commit suicide" like a lemming (**below**) when it reaches a certain point in its life. This process effectively weeds out any neurons that are not active – and therefore not essential to the survival of the whole.

Electrical stimulation within the brain seems to have a similar role in determining which connections remain, coming into play when links between neurons have to be thinned out. Those connections that are active, and are therefore generating electrical currents typical of nerve signal transmissions, survive, while those that do not create electrical signals die.

The matriarch of the elephant herd leads her group toward food, water, and safety, and away from danger. Using clues from the environment, such as scents and sights, she is the pathfinder – younger herd members follow.

In the developing brain, pioneer neuroblasts find their way between cells and membranes, following directional cues from their environment. They lay down trails that are followed by the growing fibers (axons) of other nerve cells, while the cell bodies of these fibers stay "back at base."

A shark samples waterborne substances with incredibly receptive chemosensory organs in its nasal chambers. As it turns and twists, the scents are stronger first in one nostril, then the other. In the developing human brain, the growing tips of axons do the same. Receptors on their surface detect nerve growth factor and other chemicals given off by target cells, and the tip heads for their source. Repulsion chemicals make unwanted axons turn away. As each axon lengthens, its tip homes in on the destination cell, to form a synapse, or connection, there.

Play is fun with serious consequences. As the youngster tries to put the puzzle together, fails, tries again, and finally succeeds, millions of nerve signals flash around the neurons of the nervous system. For the synapses it is a case of "use it or lose it." Well-used connections encourage further development and sophistication in that particular part of the network. Seldom-used links soon fade. This "wiring up" involves great competition and natural selection at the cellular level, as some neurons succeed, while others fail and die.

The difference in the density of connections between a newborn baby's neurons and a two year old's is incredible. In the images left, the number of neurons at each age is more or less the same, but the growth in connections comes as the infant learns.

See also

BUILDING THE BRAIN
▶ Support and protection 42/43

▶ The brain's cells 44/45

▶ The gap gallery 54/55

▶ Remaking the mind 62/63

▶ The brain plan 68/69

▶ The developing brain 70/71

▶ The cell factory 72/73

FAR HORIZONS
▶ The infinite store? 130/131

▶ First thoughts 142/143

▶ Learning to be human 144/145

▶ Milestones of the mind 146/147

STATES OF MIND
▶ The unique self 156/157

15 months   2 years

The aging brain

Our brain is not immune to the effects of old age, but the decline in performance is usually gradual.

Like other parts of the body, the brain and nerves succumb to the aging process. One problem affecting all of the nervous system is that neurons are very specialized in both shape and function. They cannot multiply to replace neurons that die, unlike the cells in skin and many other tissues; indeed, some 10,000 neurons die daily from birth.

By the late teens, the brain is still constantly adapting and re-wiring itself, but the rate at which it does so begins to slow. Learning becomes more difficult and lengthy. It is estimated that our ability to learn and remember new information could decline by 50 percent between 25 and 75 years of age. The rate of cell loss in the brain begins to increase significantly after the age of 20 and is greatly hastened by consuming too much alcohol or other drugs. On average, the brain has lost 5 percent of its weight – about 2½ ounces (70 g) – by the age of 70, although we can compensate to some extent because experience helps us to anticipate, predict, and find various types of short cuts.

Energy generation in cells produces various by-products. Damaging HO' molecules, a type of free radical, are changed by an enzyme into hydrogen peroxide (H_2O_2), which is also potentially dangerous. It can be deactivated and converted into harmless water and oxygen by another enzyme. As cells grow older, they become less efficient at producing these neutralizing enzymes, so lipid peroxidation, a type of damage that affects the lipid-containing "sandwich" that makes up cell membranes, occurs. In effect, free radicals "chew" cell membranes, especially the hydrophobic head part of the phospholipids in the membranes. This "chewing" can be disastrous for a neuron, which needs an intact membrane to conduct nerve signals and function properly.

With age, our lungs and heart become less efficient, and our blood vessels harden and narrow. The effect is fewer nutrients delivered to cells and less oxygen in the red blood cells in the capillaries supplying the brain. Hypoxia (lowered oxygen concentrations) or anoxia (lack of oxygen) may result, affecting all brain cells and causing damage and necrosis (death).

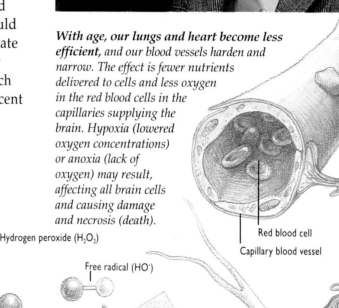

— Hydrogen peroxide (H_2O_2)

Free radical (HO')

Red blood cell

Capillary blood vessel

Astrocyte

Nucleus

Hydrogen

Nitrogen

Hydrophobic head

— Carbon

Oxygen

Phospholipid

Phosphorus

Neuron

Water (H_2O) —

Cell membrane

Dendrite

See also

BUILDING THE BRAIN
▶ Support and protection 42/43

▶ Crossing the gap 52/53

▶ Remaking the mind 62/63

▶ Recovering from damage 64/65

▶ Food for thought 66/67

▶ Linking up 74/75

SURVEYING THE MIND
▶ Brain maps 30/37

INPUTS AND OUTPUTS
▶ Keeping control 116/117

FAR HORIZONS
▶ The infinite store? 130/131

STATES OF MIND
▶ The failing mind 176/177

Such is the power of the mind that it has been suggested that if people expect to lose their memory and concentration as they grow old, then they probably will. But loss of mental function is not an inevitable part of aging. Many people in their 70s and beyond achieve wonderful feats, and make valuable contributions to their family, colleagues, and society. Indeed, the comedian George Burns was still active on his 100th birthday.

Ion channel

Calcium ion (Ca²⁺)

Mitochondrion

Axon

The opening of calcium ion channels in a neuron's cell membrane is a natural part of nerve transmission, but with increasing age, it becomes defective. Calcium ions can build up inside the cell, which can damage the energy-converting mitochondria.

HABITS AND DIETS

Some aspects of the mental diminution associated with aging may be due to habits and activities earlier in life. These include drug use, possibly even moderate "social drinking," and levels of physical impact and injury.

Diet can play its part, too. Inhabitants of the Pacific island of Guam develop brain conditions such as Alzheimer's and Parkinson's diseases many times more frequently than average, up to 100 times the normal rate. The cause may be neurotoxic chemicals eaten in the seeds of the false sago palm (**below**). This is not a true palm tree, but a non-flowering cone-bearing evergreen tree in the cycad plant group. These trees are "living fossils" with ancient origins, appearing on Earth more than 300 million years ago. They were especially abundant at the time of the dinosaurs. Perhaps those prehistoric creatures also suffered premature aging and senility after eating their leaves and seeds.

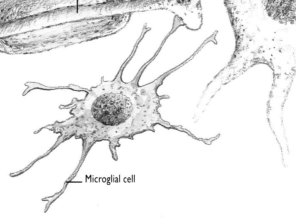

Necrosis refers to the death of cells, tissues, or parts of an organ, due to damaging external influences such as toxins, excessive heat or cold, or lack of oxygen. For example, infecting bacteria may produce necrotizing poisons (toxins). Physical injury is another cause. In the brain's case, it may be a single severe blow or many small accumulating injuries, as experienced by some boxers. The dying cells release their enzymes and contents, which disrupt the local chemical environment and produce a damaging domino effect. The debris is cleared up by small microglia – special glial cells (a type of macrophage, literally "big eater") – that operate exclusively in the brain. They have a similar role to the macrophages found elsewhere in the body.

Degenerating axon

Damage

Apoptosis is the death of single cells – including neurons – within an organ, and it may not be the result of an accident. It seems that there is a type of cellular death-wish pre-programmed by so-called suicide genes. The cell simply reaches the end of its natural life span and activates genes that instruct it to self-destruct. Unlike necrosis, apoptosis forms part of the cell's natural life cycle.

Microglial cell

Degenerating neuron

DNA

77

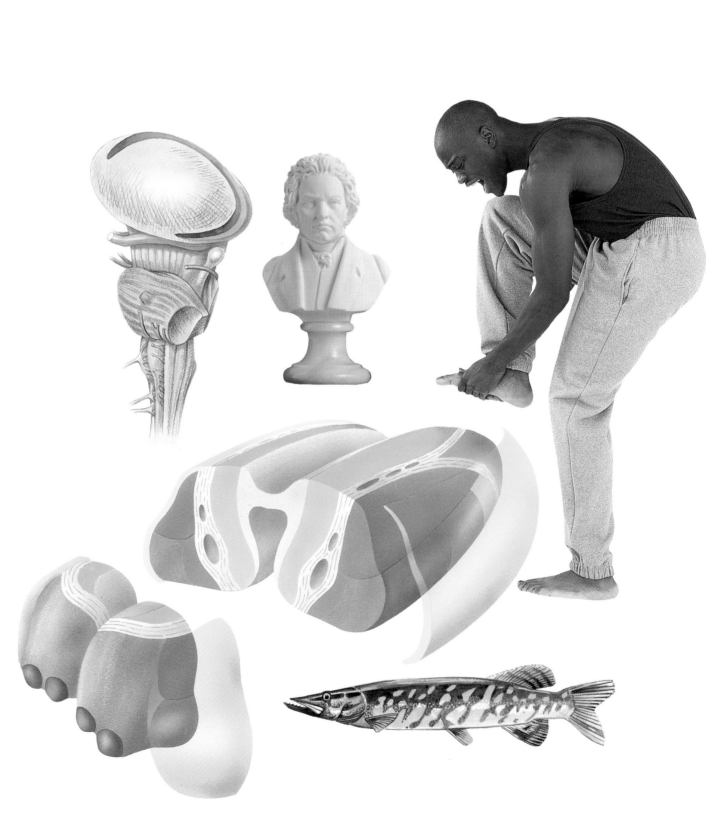

Inputs and Outputs

A s we move through the world, seeing, hearing, smelling, tasting, and touching, information ceaselessly pours into and out of the brain. This information takes the form of nerve signals, the common currency of the brain and nervous system. The story of how sound, light, and the other things we can detect with our senses are converted into signals is extraordinary enough. What must then happen to them if we are to become conscious of the world around us is even more fascinating. But the brain, along with its very own information superhighway, the spinal cord, does not merely help us to perceive the external world. It also monitors, controls, and integrates all our movements and takes care of multiple body functions – all without our having to think about them. Never at rest, it is constantly on the lookout for potential problems and has special systems devoted to focusing awareness on either external danger or internal needs.

Left (clockwise from top): the pathways of pain from foot to brain; a fishy smell; gateway to the brain – the thalamus; sound signals and active processing; music from the mind of a deaf man.
This page (top): a fitting outcome for an odor molecule; (left) a ticklish job for the feeling sense.

Body links

The brain is just one component of the nervous system, the body's command and control network.

Any part of the body which contains neurons, or nerve cells, is part of the nervous system. Some of these parts are obvious – like the bulging brain, with its 100 billion neurons, and the spinal cord, with its long tracts of signal-carrying nerve axons or fibers. But if you look more closely at these immensely complex structures, or stray into other parts of the body, it becomes more awkward to demarcate the components of the nervous system.

In order to understand any system, it helps to divide it into separate subsystems or components. But how should the nervous system be divided? Should structural or anatomical criteria, such as shape and size, be applied; or can functional or physiological aspects, such as which neurons link together, be used? You may as well take the whole category of road vehicles and try to define a car solely on the basis of length, or color, or noise.

In fact, a "car" is defined by several different sets of criteria which overlap, and the nervous system can also be subdivided in different yet overlapping ways, each with its own terminology. For example, an afferent, or sensory, neuron conveys signals from a sensory part of some kind, such as the skin or eyes, toward the brain. An efferent, or motor, neuron carries nerve signals away from the brain, usually to a muscle. Yet many of the stringlike "nerves" winding through the body contain the axons of both sensory and motor neurons. Similarly, a closeknit lump or group of neuron cell bodies inside the brain or spinal cord is known as a nucleus, whereas the same type of structure in another part of the nervous system is a ganglion.

There are also large groups of nerves – for example, cranial, cervical, sacral – many of whose individual nerves have a name. Often a name is derived from a nearby bone, such as the tibial nerve which runs along the tibia (shinbone).

CRANIAL NERVES: BRANCHES OF THE HEAD OFFICE

The unique feature of cranial nerves is that they branch directly from the brain, rather than from the spinal cord. There are 24 in 12 pairs, linking the brain to parts mainly in the head and neck, but also in the trunk. They are known by Roman numerals, and their main branches also have functional or anatomical names. For example, cranial nerve VII is the facial nerve. Some nerves are purely sensory, others are mainly motor (muscle control) nerves, and still others are mixed. The colored parts of the face (**right**) show the different areas from which sensory messages are carried to the brain by three branches of cranial nerve V, the trigeminal. The network of lines shows some of the motor branches of the facial nerve to the same areas.

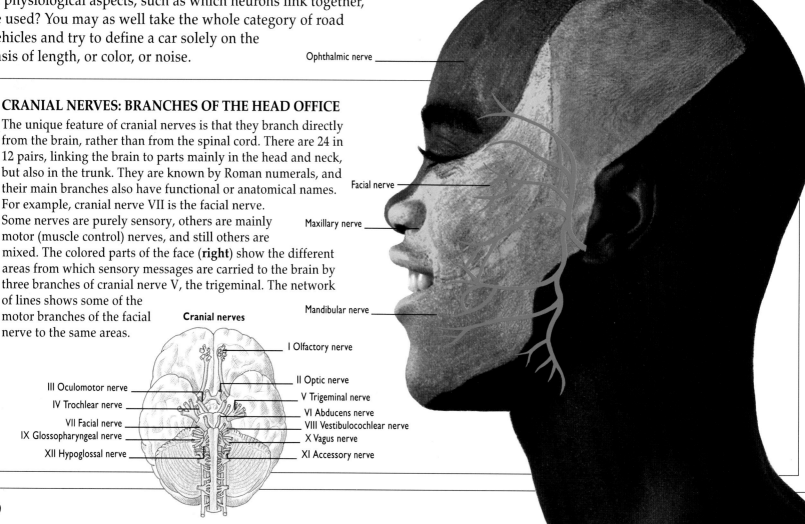

Ophthalmic nerve

Facial nerve

Maxillary nerve

Mandibular nerve

Cranial nerves

I Olfactory nerve
III Oculomotor nerve
IV Trochlear nerve
VII Facial nerve
IX Glossopharyngeal nerve
XII Hypoglossal nerve
II Optic nerve
V Trigeminal nerve
VI Abducens nerve
VIII Vestibulocochlear nerve
X Vagus nerve
XI Accessory nerve

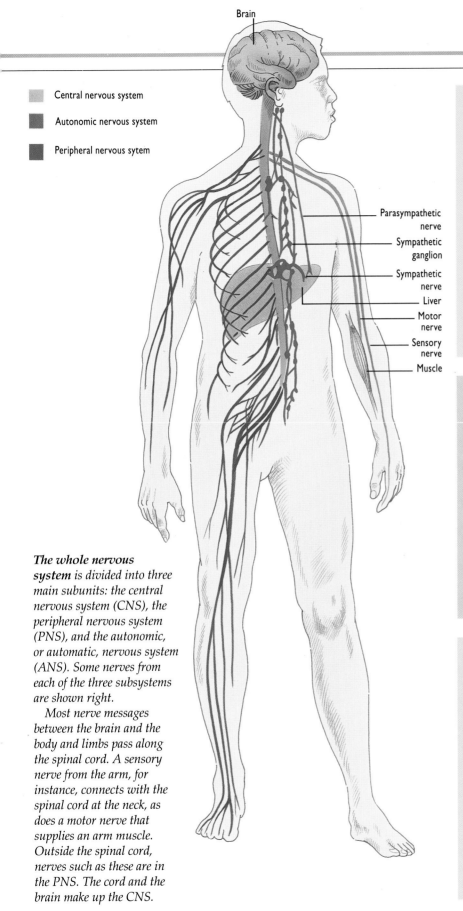

Brain

Central nervous system

Autonomic nervous system

Peripheral nervous sytem

Parasympathetic nerve

Sympathetic ganglion

Sympathetic nerve

Liver

Motor nerve

Sensory nerve

Muscle

The whole nervous system is divided into three main subunits: the central nervous system (CNS), the peripheral nervous system (PNS), and the autonomic, or automatic, nervous system (ANS). Some nerves from each of the three subsystems are shown right.

Most nerve messages between the brain and the body and limbs pass along the spinal cord. A sensory nerve from the arm, for instance, connects with the spinal cord at the neck, as does a motor nerve that supplies an arm muscle. Outside the spinal cord, nerves such as these are in the PNS. The cord and the brain make up the CNS.

CENTRAL NERVOUS SYSTEM (CNS)

Together, the brain and spinal cord form the central nervous system. They are central in terms of their position within the head and body, and they are, of course, central to the functioning of the whole nervous system and so the entire body. They are the only parts of the nervous system that are well protected by surrounding bones, the brain inside the cranial bones of the skull, and the spinal cord within the linked vertebrae of the spinal column (backbone). The spinal cord, containing millions of nerve fibers, is the body's biggest nerve; the brain has the body's largest concentration of neurons.

PERIPHERAL NERVOUS SYSTEM (PNS)

The peripheral nerves are pale, whitish, shiny, cordlike structures snaking between the body's organs. They form the links between the CNS and other body parts, from head to toe. In addition to the 12 pairs of cranial nerves, there are 31 pairs of spinal nerves that branch from along the spinal cord, at junctions known as spinal nerve roots. Some of the spinal nerves branch from the cord, join together and interweave, then branch again. Each of these networks, which look like aerial views of complex road junctions, is known as a nerve plexus.

AUTONOMIC NERVOUS SYSTEM (ANS)

The ANS controls and coordinates the body's "automatic" functions – basic life processes such as heartbeat, digestion, excretion and, to an extent, breathing. There are two major subdivisions with largely complementary roles: the sympathetic ANS generally prepares the body for action; the parasympathetic ANS has a more "peacetime" function, thus it encourages the body's digestive processes. For instance, the sympathetic system signals the liver to release glucose to give energy to the skeletal muscles. The parasympathetic changes priorities to promote the liver's digestive functions.

See also

INPUTS AND OUTPUTS
▶ The long junction 82/83

▶ On the move 84/85

▶ Guided motion 86/87

▶ Levels of seeing 92/93

▶ Sensing sound 96/97

▶ Pain pathways 102/103

▶ A matter of taste 106/107

▶ On the scent 108/109

▶ The big picture 114/115

▶ Keeping control 116/117

SURVEYING THE MIND
▶ Discovering the brain 14/15

▶ Brain maps 30/37

FAR HORIZONS
▶ Parallel minds 138/139

The long junction

Like a major highway, the spinal cord carries nerve signals to the brain from the body and vice versa.

Just as an experienced servant fetches and carries for his long-time master, so the spinal cord conveys nerve signals to and fro between the brain and the body – or, more precisely, between the brain and the chest, abdomen, and limbs. (The head and face link to the brain by cranial nerves.) But the cord is no mere passive servant, simply conveying messages unaltered. It carries out its own relaying and processing of nerve signals. It also works independently of the brain when it activates various body reflexes, such as pulling your hand away from something hot. The vital nature of the cord becomes strikingly clear when it is damaged, by either disease or injury. Depending on the site and severity of the damage, the effects can range from tingling and weakness in the extremities to total paralysis, incontinence, loss of reflexes, and loss of sensation.

Structurally, the spinal cord is effectively an extension of the brain: in embryonic development, both originate from the same set of cells, the neural tube. There is no obvious external joining between them, only a smooth merger. So the top of the cord is usually taken to be at the level of inside the foramen magnum, the large downward-facing hole in the base of the skull. A long tube, the vertebral canal, protects and encloses

At the railroad interchange and classification yard, *trains are routed through a complex maze of tracks and points, on to their correct destination. The spinal cord directs nerve signals in a similar manner, through its network of neurons and their dendrites and axons. Signals pass from the body to the brain along major ascending spinal tracts, or "up lines," and from the brain back into the body along the "down lines," or descending spinal tracts.*

REFLEX ACTION

Stroke the outer part of the sole of a new baby's foot, and its toes curl. This is the Babinski reflex, one of many such reflexes – quick automatic responses to certain stimuli – in which signals from a sense organ pass along sensory nerves to the spinal cord. There the signals are processed and return ones go via motor nerves to move the muscles. Other signals go up the cord to the brain, which becomes aware of what has happened – but too late to affect the reflex. New babies are bundles of reflexes, each activating a simple survival action such as sucking, swallowing, or emptying the bowel and bladder. With age, children learn to control many of these responses.

the cord. It is formed by the row of holes inside the closely joined vertebrae of the spinal column, or backbone. In an adult, the spinal cord is usually about 18 inches (45 cm) long and slightly thinner than an index finger. It does not extend down the whole spine to the base of the spinal column, but tapers rapidly at the level of the first lumbar vertebra, just above the waistline, into a nerveless strip of fibrous tissue known as the filum terminale, which is attached to the coccyx. The vertebral canal continues below the first lumbar vertebra, but contains nerves that have branched from the cord.

Membranes and fluid – more extensions from the brain above – surround and cushion the cord. There are three layers, or meninges, and cerebrospinal fluid (CSF). The cord also has a tiny tube, the central canal, along its length; this contains CSF, too, which is continuous with the CSF in the ventricles inside the brain. In its nerve layout, the cord is the "reverse" of the cerebral cortex. It has white matter – mainly myelinated nerve fibers or axons – on the outside; inside is the gray matter of connections and nerve cell bodies, such as processing interneurons, which forms a cross-sectional H or butterfly shape.

See *also*

INPUTS AND OUTPUTS
▶ Body links
80/81

▶ On the move
84/85

▶ Guided motion
86/87

▶ Sense of touch
100/101

▶ Pain pathways
102/103

▶ Keeping control
116/117

SURVEYING THE MIND
▶ Brain maps
30/37

BUILDING THE BRAIN
▶ Support and protection
42/43

▶ Sending signals
50/51

FAR HORIZONS
▶ First thoughts
142/143

Cortex

Descending nerve

Collateral

Midbrain

Cerebrum

Pons

Cerebellum

Upper medulla

Cervical nerves
8 pairs

Nerves to arm

Lower medulla

Thoracic nerves
12 pairs

Lumbar nerves
5 pairs

Sacral nerves
5 pairs

Nerves to leg

Spinal tracts, or pathways, are bundles of nerve fibers, or axons, in the white matter of the spinal cord. There are two major types of tracts, ascending and descending, named according to their positions in the cord and the brain parts that they link to.

Ascending tracts are sensory, taking nerve signals concerned with sensation up to the brain. The posterior and anterior spinocerebellar tracts carry sensory messages about the position of joints and muscles to the brain stem, for mainly unconscious assessment of body position and posture (known as proprioception). The lateral spinothalamic tract conveys signals from the body about pain and temperature; the anterior spinothalamic tract carries information about light or coarse touch. The spinoreticulothalamic tract is grouped with these and also concerns pain. The lemniscal tracts in the back of the cord are evolutionarily newer pathways for more discriminatory messages about fine, detailed touch, which go from the skin to the somatosensory cortex, the brain's touch center.

Descending pathways   **Ascending pathways**

Lateral corticospinal tract

Lemniscal tracts

Rubrospinal tract

Gray matter

White matter

Cell body of sensory neuron

Reticulospinal tracts

Ganglion   Interneuron

Spinocerebellar tracts

Incoming sensory nerve

Outgoing motor nerve

Vestibulospinal tract

Anterior corticospinal tract

Spinothalamic tracts

Spinal nerves branch from the spinal cord in pairs between adjacent vertebrae. Their names reflect the groupings of these vertebrae: cervical (C), or neck; thoracic (T), or chest; lumbar (L), or back; and sacral (S), or lower back. If you decide to move your right fingers, nerve signals pass from the left motor cortex down to the pons and medulla. As this tapers into the spinal cord, the signals cross to the right side and continue down the lateral corticospinal tract, passing out along right spinal nerves C7 to T2. Motor nerves leave the spinal cord at its front, carrying the signal down to the finger. Any returning sensory information enters the cord at its rear.

Descending pathways in the spinal cord carry motor nerve signals from the brain on their way to the body's muscles to make movements. The lateral (pyramidal) corticospinal tract is concerned with detailed movements of the extremities and in particular – together with the rubrospinal tract – with finger manipulation. The lateral (medullary) and medial (pontine) reticulospinal tracts convey messages to many muscles in the torso, shoulders, and hips, for the seldom-noticed movements that maintain or change posture, equilibrium, and balance. The signals going along the vestibulospinal tract are similar, and they also go to the limbs for fine balancing movements. The anterior (pyramidal) corticospinal tract conveys nerve signals to the muscles of the neck and torso, for bending, twisting, and other voluntary movements.

The various tracts in the spinal cord are primarily named after the areas where they start and finish. "Spino-" refers to the gray matter in the spinal cord, for instance, "thalamic" the thalamus, "cortico" the cerebral cortex, and so on. So the spinothalamic tract starts in the cord's gray matter and ends in the thalamus.

On the move

Like a puppeteer pulling strings, the brain sends signals that make the muscles contract and move the body.

To obtain the comforts of life, the brain sends out messages as nerve signals to the various systems of the body, such as the circulatory system for nourishment via the blood and the musculoskeletal system for physical motion. The body has about 640 skeletal muscles, ranging in size from the huge slab of the gluteus maximus in the buttock and upper rear thigh to the threadlike stapedius deep in the ear. Skeletal muscles work by getting shorter, or contracting. Most of them are attached by their tapering, ropelike ends (tendons) to bones. As the muscle contracts, its tendon pulls on the bone and moves it. The bone, being strong and rigid, acts as a firm inner support and moves that whole part of the body. Body movements get food, find shelter, and avoid danger – helping the body and brain survive.

The skeletal muscles do not work alone. In effect, the muscular body is a robot, controlled by the brain. This sends nerve signals along motor nerves to all muscles, telling them when and how to contract and make a movement, or to relax and go floppy, so they do not oppose the work of other muscles. The more signals per second, the more the muscle contracts. Even the simplest movement such as waving goodbye is a massive cooperative effort by dozens of muscles working together in coordinated teams with split-second timing, each muscle controlled by thousands of nerve signals every second.

Several parts of the central nervous system – the brain and spinal cord – are dedicated to or partly involved in this process. A complex system of continuous monitoring and feedback makes sure movements are smooth and coordinated – and exactly what the brain requires.

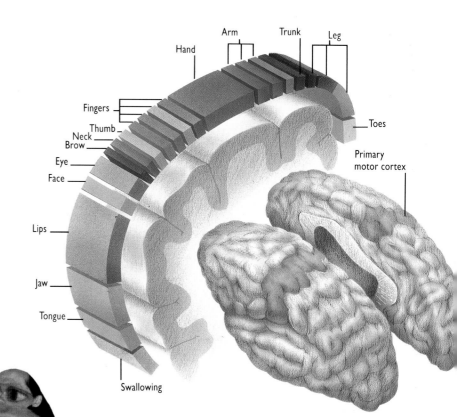

The chief executives in the movement process are two strips of gray matter over the top of the brain, just under where you might wear a stretchy hairband or a set of headphones. They are the left and right motor cortices – the left one controlling the muscles on the body's right side, and the right one those on the body's left side. The different sets of muscles in each part of the body have their own patch of motor cortex, resulting in a "strip map" of the body on the brain, from face and head on the outer (temple) side, to legs, feet, and toes on the inner (midline) side. The larger the area of the cortex, the greater the precision of control. So a body part like the hand, which requires considerable precision to perform tasks such as using tools, will have a far greater area of cortex devoted to it than, say, the trunk, whose movements can be fairly imprecise.

Each body part of the motor homunculus (little man) is sized in proportion to the area of the brain's motor cortex that controls the muscles there. This indicates not the size and power of a muscle, but how precisely it can be controlled. Parts which can be moved very accurately with fine control, such as the lips, tongue, and fingers, are largest.

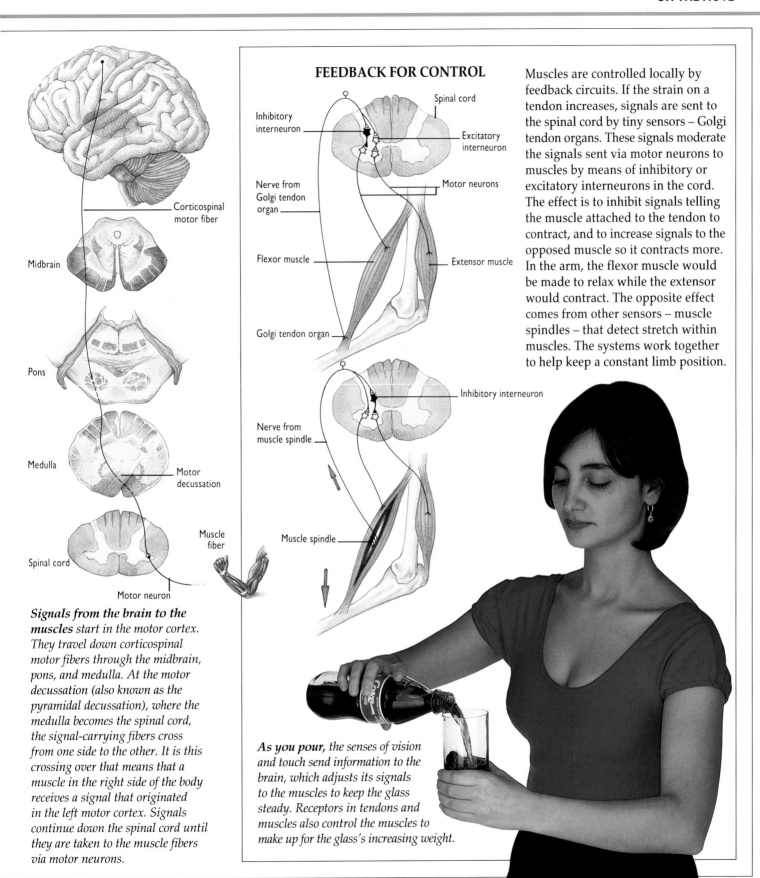

FEEDBACK FOR CONTROL

Inhibitory interneuron

Spinal cord

Excitatory interneuron

Nerve from Golgi tendon organ

Motor neurons

Flexor muscle

Extensor muscle

Golgi tendon organ

Inhibitory interneuron

Nerve from muscle spindle

Muscle spindle

Corticospinal motor fiber

Midbrain

Pons

Medulla

Motor decussation

Spinal cord

Muscle fiber

Motor neuron

Muscles are controlled locally by feedback circuits. If the strain on a tendon increases, signals are sent to the spinal cord by tiny sensors – Golgi tendon organs. These signals moderate the signals sent via motor neurons to muscles by means of inhibitory or excitatory interneurons in the cord. The effect is to inhibit signals telling the muscle attached to the tendon to contract, and to increase signals to the opposed muscle so it contracts more. In the arm, the flexor muscle would be made to relax while the extensor would contract. The opposite effect comes from other sensors – muscle spindles – that detect stretch within muscles. The systems work together to help keep a constant limb position.

Signals from the brain to the muscles start in the motor cortex. They travel down corticospinal motor fibers through the midbrain, pons, and medulla. At the motor decussation (also known as the pyramidal decussation), where the medulla becomes the spinal cord, the signal-carrying fibers cross from one side to the other. It is this crossing over that means that a muscle in the right side of the body receives a signal that originated in the left motor cortex. Signals continue down the spinal cord until they are taken to the muscle fibers via motor neurons.

As you pour, the senses of vision and touch send information to the brain, which adjusts its signals to the muscles to keep the glass steady. Receptors in tendons and muscles also control the muscles to make up for the glass's increasing weight.

See also

INPUTS AND OUTPUTS
▶ Body links 80/81

▶ The long junction 82/83

▶ Guided motion 86/87

▶ Staying upright 112/113

▶ The big picture 114/115

▶ Keeping control 116/117

▶ Survival sense 122/123

SURVEYING THE MIND
▶ Inside the mind 20/21

▶ Brain maps 30/37

BUILDING THE BRAIN
▶ Sending signals 50/51

FAR HORIZONS
▶ Parallel minds 138/139

Guided motion

Decide to make a move, and parts of the brain and spinal cord frantically send signals back and forth.

Imagine (or recall) practicing a new and complex activity such as serving a tennis ball. It is painfully slow and awkward at first. Each muscle seems to need individual attention as you try to move the limbs, hands, and fingers in the correct way, while retaining posture and balance. The first few times you might miss the ball or even fall over. But gradually the motor skills improve. You guide your movements more accurately, using your senses – watching the ball and racquet, feeling with your hand and fingertips, and listening for the ball's ping on the racquet strings. This sensory-guided motion is steered by feedback from the senses.

After a few more hours' practice, the serve seems to happen almost automatically, without your really thinking. It is now a subconscious motor skill, a series of pathways, connections, and sequences between neurons in the brain and spinal cord. The serving movements are now well learned, faster, and more powerful – a ballistic motion that, once initiated and in progress, leaves too little time to adjust and fine-tune it. It is literally hit or miss.

A tennis player's first serve is usually fast and hard. The brain operates the muscles in a rapid sequence that has been programmed in during practice. This type of action, where there is no time for correction, is termed ballistic. The second serve is often slower and more controlled. The player adjusts the speed, force, and angle of the racquet using proprioceptive feedback from sensors in muscles, tendons, and joints.

A deliberate sensory-guided movement involves complex, ongoing messages being sent between several brain and spinal cord regions. The decision to move is usually prompted by sensory information that reaches the primary planning region, the association cortex. This sends signals to basal ganglia and to the cerebellum where movement patterns, or motor commands, are evoked and sent to the motor cortex via the thalamus. The motor cortex plays the pivotal role between receiving a motor command and executing it. It sends signals to the motor centers in the spinal cord both directly and via the brain stem. The motor centers consist of spinal interneurons and motor neurons, which send signals to make the muscles move. The motor cortex also sends data back to the cerebellum and the basal ganglia, and the various components receive feedback as sensory inputs cause ongoing corrections to the movement being made.

Planning

Programs

Association cortex

Sensory inputs

Cerebellum

Thalamus

Motor cortex

Basal ganglia

Maintenance of posture and execution of movement

Brain stem

Spinal interneurons

Motor neurons

Muscles

See also

INPUTS AND OUTPUTS
▶ Body links
80/81

▶ The long junction
82/83

▶ On the move
84/85

▶ Staying upright
112/113

SURVEYING THE MIND
▶ Brain maps
30/37

FAR HORIZONS
▶ Parallel minds
138/139

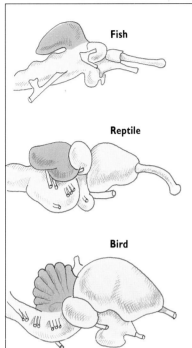
Fish

Reptile

Bird

Cat

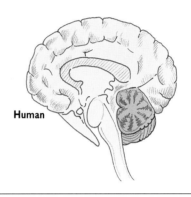
Human

COMPARING LITTLE BRAINS

 The cerebellum, or "little brain," makes up about one-tenth of the volume of a human brain. In some fish, it is nine-tenths. In animals such as fish and reptiles, it is the prime controller of body movements. In birds, and to a greater extent mammals, above all humans, the "higher" brain areas of the motor cortex take over control of voluntary movements. But the cerebellum still plays a role in precise, balanced coordination of the muscles, once a motion has begun. It receives nerve signals about the central motor program from the cortex, and signals from microsensors in muscles, tendons, and joints, about the results of a movement as it happens. The cerebellum compares the instructions and results, assesses how they match, and outputs signals to the motor cortex to modify and fine-tune the program.

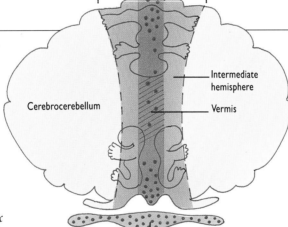
Spinocerebellum

Intermediate hemisphere

Cerebrocerebellum

Vermis

Vestibulocerebellum

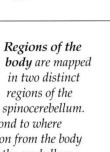
Regions of the body *are mapped in two distinct regions of the spinocerebellum.* The maps correspond to where sensory information from the body regions arrives in the cerebellum.

Cerebrocerebellar *regions (yellow) send signals to motor cortex areas and have a role in motor planning.*

Ongoing movements, *such as walking, are controlled by the spinocerebellum (which includes the vermis).*

Signals from the inner ear *concerning head movements and gravity go to the vestibulocerebellum. It controls and coordinates head–hair eye movements, so you can watch a speeding object pass by, moving your eyes and twisting your neck in a coordinated manner to keep it in view.*

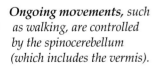
The various cerebellar regions *receive and send signals to different parts of the brain and nervous system. The colors and markings (the dots, and the diagonal lines in the center of the vermis) on the diagram of the cerebellum above indicate which signals the cerebellar regions receive and the functions that they control. Thus, the cerebrocerebellum (yellow) sends signals to and receives them from the cerebral cortex and other regions including the pons (**left**).*

The central region of the vermis *receives inputs from both the visual and the auditory systems. These incoming signals assist it in controlling muscles in the trunk and other regions, and in controlling and coordinating the continuing execution of movements.*

The vestibulocerebellum and vermis *use signals from the vestibular system to control body balance, and the muscle tensions and movements that maintain it. This means you can sit, stand and move without having to think about your posture and balance.*

The spinocerebellum *receives inputs from the spinal cord and other regions. It is responsible, among other things, for maintenance of muscle tone and the elimination of tremors.*

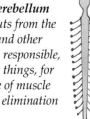

Seeing the light

The story of how we see the world about us starts in the eye, where light is turned into coded nerve signals.

Humans, like their primate relations the monkeys and apes, are intensely visual animals. The eyes send millions of nerve signals every second along the optic nerves to the visual pathways in the brain, for analysis, interpretation, and recording. The nerve signals represent energy transformed, or transduced, from the energy in light rays. This happens in the 1-inch (2.5-cm) diameter "biological camera," specifically in its light-sensitive layer, the retina, which is about the size and thickness of a postage stamp and lines the inner rear two-thirds of the eyeball. Around it is the blood-rich, eye-nourishing layer known as the choroid, surrounded in turn by the tough white outer sheath of the eyeball, the sclera. Within the retina is the vitreous humor, a clear jelly that gives the eyeball shape and firmness without obstructing light rays.

Rays of light that fall on the retina have already been altered to produce a clear, sharp, unblurred image there that is of suitable brightness or intensity. This is achieved by optical structures at the front of the eye. Light passes from the object being observed into the eye through its clear front, the cornea, whose domed shape does most of the focusing of the rays. It then passes through the pupil, an adjustable hole in the middle of the iris. This is a ring-shaped system of muscles which give the eye its color. Like the aperture on a camera, the iris makes the pupil smaller in bright light conditions – to restrict the amount of light entering, which might damage the retina if too intense – and larger if it is dim, allowing the optimum amount of light in to give the best possible image.

Next, the light rays pass through the lens, which can be pulled thin or left fat by another ring of muscles around it, the ciliary muscles. This "fine-tunes" the focusing, according to whether the rays come from a near or distant object, to throw a sharp image onto the retina.

Mitochondrion

Nucleus

Synaptic terminal
Synaptic vesicle

Dendrites of bipolar and horizontal cells

Receptors in retina

Rods

Cones

Inside the eye, light rays pass through the cornea and then the pupil in the center of the iris. The fixed cornea and adjustable lens bend light rays to focus a clear image onto the retina. As with an equivalent man-made lens, the image is inverted. But since this is the case from birth, we never know any different, and so it is not a problem.

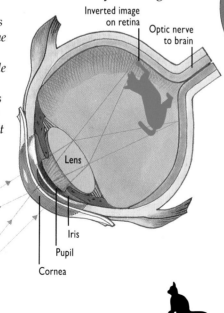

Inverted image on retina

Optic nerve to brain

Lens

Iris

Pupil

Cornea

The retina *has two types of light-sensitive cells – rods and cones (named after their shapes). The 125 million rods detect shades of black and white. The 5 to 7 million cones detect color and fall into three types, each of which is most sensitive to one of the primary colors of light: red, blue, and green. Most cones are in the center of the retina, especially in the fovea – a rod-free area where vision is sharpest. Rods, and some cones, are found in the rest of the retina.*

Each rod cell (above) *is about 150–200 micrometers ($\frac{3}{500}$–$\frac{1}{125}$ inch) long. At one end, next to a layer of pigment cells and facing the outside of the eyeball, it has a stack of disks studded with the chemicals that play a part in transducing light energy. Toward its other end, it has a nucleus and other standard cell components such as mitochondria. At its base, it synapses (connects) with dendrites of intermediary cells which link it to retinal ganglion cells. These send nerve signals to the brain, and their axons (message carrying fibers) form the optic nerve. Each retina has a blind spot, a receptor-free area where all the axons leave the eye.*

To find the blind spot *of your right eye, close your left eye and look at the black dog. Adjust the distance of the book until the cat vanishes. Repeat for the left eye, but this time close your right eye and look at the cat. The dog should disappear, but the lines seem continuous. The brain actively fills in with what it "thinks" should be there, in this case linking the vertical lines.*

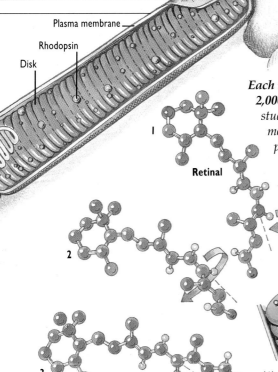

Plasma membrane
Rhodopsin
Disk
Retinal

1
2
3

Before light can reach rods and cones at the rear layer of the retina, it travels through several other layers. These are made up of tiny blood vessels and the so-called neural cells of the retina – bipolar, horizontal, and ganglion cells – and are effectively transparent. Behind rods and cones is a layer of pigment cells which absorb any stray light and prevent it from reflecting back to the retina.

Each rod cell contains some 2,000 stacked disks, which are studded with up to 100 million molecules of the light-sensitive pigment rhodopsin. Each rhodopsin molecule has two parts – the opsin protein and the light-absorbing substance retinal, derived from vitamin A, which can exist in several structural forms, or isomers. Before light hits it, the retinal is in the form 11-cis-retinal (**1**). When a particle or packet of light energy – a photon – hits rhodopsin, it makes one end of 11-cis-retinal twist around (**2**) to form another isomer, all-trans-retinal (**3**). This changes the configuration of the opsin protein, too, converting the whole molecule from rhodopsin into metarhodopsin II – in the space of only one-thousandth of a second. Giving the retinal the energy to change shape is the only part light plays in this process.

The transformation of rhodopsin to metarhodopsin II triggers a series of chemical reactions within the rod. First each metarhodopsin II molecule activates hundreds of molecules of the protein transducin. Each of these in turn activates a type of enzyme known as a phosphodiesterase, which alters the structure of thousands of molecules of the neurotransmitter that stimulates cells in the retina – cyclic guanosine monophosphate (cGMP). This reduces levels of cGMP in the rod, which in darkness are high.

As the concentration of cGMP in the rod cell falls, channels which allow sodium ions to flow through its membrane close. This is because cGMP's function is to keep these channels – found in the rod cell's membrane next to the region containing the disks – open. Thus when no light falls on the retina, the channels allow a flow of positively charged sodium ions into the cell – known as the dark current – to counterbalance the diffusion of positive potassium ions out of the cell, making the inside slightly negative. When light hits the cell, sodium entry reduces, but potassium exit continues, so the cell's interior becomes more negative, or hyperpolarizes.

Hyperpolarization of the rod cell reduces the release of neurotransmitters from synaptic vesicles in its synaptic terminal. This results in signals being sent to the brain. Meanwhile, the chemicals used in the process are recycled: all-trans-retinal reverts to 11-cis-retinal, and the sodium channels reopen. It all happens within one-fifth of a second of the original photon first reaching the rod cell. The process in cones is similar to that in rods, but cones – which work best in bright light – require about 50 times more light to activate them to the same extent.

See also

INPUTS AND OUTPUTS
▶ Contrasting views
90/91

▶ Levels of seeing
92/93

▶ Active vision
94/95

▶ The big picture
114/115

▶ Survival sense
122/123

SURVEYING THE MIND
▶ Comparing brains
18/19

BUILDING THE BRAIN
▶ The electric cell
46/47

▶ To fire or not to fire?
48/49

▶ Crossing the gap
52/53

▶ On or off
56/57

▶ Unlocking the gate
58/59

MORE THAN MEETS THE EYE

The human eye has the structure typical of any mammal's eye, which in turn is very similar to the eye of any vertebrate – bird, reptile, amphibian, or fish. However, there are many other designs of eye, or image-forming photosensory organ, in the animal kingdom.

Insect eyes – like that of this horsefly, *Tabanis* sp. – are compound, made of hundreds, or even thousands, of closely packed light-sensitive elements known as ommatidia. Each of these – in effect a separate eye – detects light rays from only a tiny part of the scene, but the many units may combine to produce an overall image in much the same way as a mosaic is built up from many small elements. The "brain" of the insect integrates the multiple images so that although a fly cannot detect much detail with such an eye, it can pick up objects moving across its visual field fairly well.

Contrasting views

The processing of signals in the visual system starts with complex layers of nerve cells in the retina.

An orbiting space telescope takes a photograph deep in space. On the TV monitor screen back on Earth, it appears as a chaotic swirl of dark and light blobs. But the image-enhancement computer sharpens differences, or contrasts, and defines clear shapes from the gloom. The retina of the eye works in a similar way. Numerous light-sensitive photoreceptor cells – rods and cones – form its outer layer. Within this layer are more hugely complex layers of interconnected neurons.

Signals from possibly hundreds of rods and cones feed into a dozen or so bipolar cells, which then send signals to one ganglion cell. The system constitutes the receptive field of that ganglion cell and carries out what computer scientists refer to as data compression. The ganglion cell's firing rate is governed by the summation and integration of the signals from all of its photoreceptors, and is thus an indicator of the amount and distribution of light falling on its receptive field.

These data compression–convergence routes are the direct pathways for visual information, but there are also lateral pathways involving horizontal and amacrine cells. As a result of these interacting pathways, signals generated by about 130 million rods and cones are pre-processed, image-enhanced, and data-compressed in the optoneural network of the retina, producing signals that are sent from the 1 million ganglion cells in the retina to the brain. These pass as nerve impulses along the ganglion cell fibers (axons), which form the optic nerve.

An eagle could spot a rabbit at 3 miles (5 km), but we would struggle to identify it at just ³⁄₅ mile (1 km) away. An eagle's light-sensitive rods and cones are packed five times more densely in the retina than ours, averaging about 6.5 million per ¹⁄₁₀₀ sq. inch. Our retina's fovea, where cones are most concentrated to see in greatest detail, is a flattish disk. In a hunting bird, it is pit-shaped, which has a magnifying effect.

COLOR-CHALLENGED

A young man applied to join the air force, but after tests was turned down because his color vision was defective. He was not color-blind, a rare state in which a person sees only in monochrome – like a black-and-white movie – but he had problems telling reds from greens. Thus, he would not have been able to distinguish the umbrella in the picture on the right. This common condition is usually inherited, and it almost always affects males. Overall, color vision defects affect about 1 in 12 men and 1 in 200 women; red–green defects are most common. Some affected people never know, unless they are tested. They assume other people see colors in the way they do; they learn to distinguish delicate differences between similar hues, especially in bright light.

Just three colors are used to make the image above, yet the different combinations make it seem as if there are more. This is because our perception of a color is affected by its context, that is, by the colors that are adjacent to it.

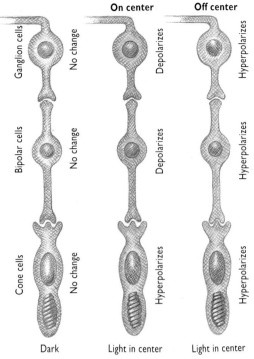

On center | Off center

Ganglion cells — No change | Depolarizes | Hyperpolarizes

Bipolar cells — No change | Depolarizes | Hyperpolarizes

Cone cells — No change | Hyperpolarizes | Hyperpolarizes

Dark | Light in center | Light in center

Ganglion cells

Bipolar cells

Cone cells

Horizontal cell

On center receptive fields

Cell depolarizes

Cell hyperpolarizes

No change in polarization

Light

Light-sensitive photoreceptor cells – rods and cones – packed into the retina do not send signals directly to the brain. Instead information is sent via bipolar cells to ganglion cells, which send signals to the brain along the optic nerve.

When illuminated, a rod or (as in this diagram) a cone hyperpolarizes. What then happens at the bipolar cell stage depends largely on the type of receptive field of the relevant ganglion cell and on the cone's position in the field. Each retinal ganglion cell has a receptive field made up of an outer ring, or surround, and an inner circle, or center. There are two main types of receptive fields: on- and off-center. In the on-center field, illuminating the cones in the middle makes them hyperpolarize, which causes depolarization in both the bipolar and ganglion cells above. This makes the ganglion cell fire faster than its normal "dark" background rate; illuminating cones in the surround makes the ganglion cell fire more slowly than normal.

When a cone in the middle of an off-center field hyperpolarizes, the bipolar cell above it also hyperpolarizes, causing the ganglion cell above to hyperpolarize, too, which reduces its rate of firing. Illuminating the cones in the surround of an off-center field makes the ganglion cell fire faster than normal. The brain interprets the patterns of changes in firing rates of the ganglion cells across the retina in the process of visual perception.

An additional feature of the system is the lateral inhibition of ganglion cells. This is carried out by horizontal cells, which attach to adjacent rod or cone cells, and amacrine cells (not shown) which link to neighboring ganglion cells. Horizontal cells pass inhibitory signals from hyperpolarized cones to adjacent cones. Here, in this highly simplified diagram, a row of horizontal cells connects a row of illuminated, hyperpolarized cones with a row in the dark. The horizontal cells make the non-illuminated cones depolarize, and this makes the bipolar cells above them hyperpolarize, lowering the rate of firing in the ganglion cells. The effect is to increase differences and sharpen contrasts, one part of retinal image enhancement.

See also

INPUTS AND OUTPUTS
▶ Body links 80/81

▶ Seeing the light 88/89

▶ Levels of seeing 92/93

▶ Active vision 94/95

▶ The big picture 114/115

▶ Survival sense 122/123

SURVEYING THE MIND
▶ Discovering the brain 14/15

▶ Brain maps 30/37

BUILDING THE BRAIN
▶ The brain's cells 44/45

▶ To fire or not to fire? 48/49

▶ On or off 56/57

FAR HORIZONS
▶ First thoughts 142/143

Levels of seeing

More is known about how the brain processes nerve signals from the eyes than those from any other sense.

Nerve signals from the 1 million ganglion cells in the retina of each eye pass along the optic nerve to a half-crossover junction – the optic chiasma. The signals continue along the optic tracts to paired parts of the thalamus known as lateral geniculate nuclei, or LGNs. They then continue along fan-shaped optic radiations to their main destination, the visual cortex of each occipital lobe. These are sited at the lower central back of the cerebrum, just inside the partially protruding lump of the lower rear skull.

The visual cortices are sight centers concerned with decoding and analyzing the nerve signals from the retinal ganglion cells. Each region of visual cortex has a number, so the primary visual cortex, the main reception area for visual signals, is V1. It is effectively a "copy map" of the retina. The activities of V1's vast patchwork of neurons represent a mosaic

A ballet dancer pirouettes and leaps gracefully, yet a few people would see a series of staccato images. This is because of damage to parts of the visual cortex, which can disrupt some aspects of vision while leaving others undisturbed. Thus shapes, outlines, colors, and contours may be unaffected, but motion is perceived as disjointed.

SEEING WITHOUT KNOWING

A man who had partial blindness after a head injury underwent tests on his visual field – the area that he could still see. The doctor flashed lights across various parts of the normal visual field. The man saw and pointed to them, until the lights were in the now-blind area, when he said he could no longer see them. But when coaxed to guess where the lights were, he pointed to them in a repeated and reliable way, with a surprising degree of accuracy – yet he still denied seeing anything at all.

This condition is blindsight. The eyes and nerve pathways to the visual cortex function normally, yet the person firmly states that, in his conscious awareness, he sees nothing. But the images must register at some lower level, since the brain reacts by instructing muscles to make the arm point to them. Blindsight is associated with damage – usually

severe – to the main primary visual cortex (region V1) at the lower rear of the brain. This is the chief reception site for nerve signals coming from the eyes via the major relay stations of the LGNs (lateral geniculate nuclei) in the thalamus. Possibly V1 is involved in passing signals on to other visual regions from where signals spread through the brain and enter consciousness; this does not happen in blindsight.

However, nerve pathways may be operating that send signals to other parts of the visual cortex, bypassing V1. These are thought to be from the superior colliculi and from the LGNs, which connect with region V4 from where further signals are sent to the brain's motor system. Thus the body can point to an image, but no signals reach conscious perception, so the person denies the image's existence.

reflecting the pattern of signals sent in by the retinal ganglion cells. Around it in the secondary visual cortex are regions V2, V3, and so on. They sort, and to an extent separately process, the various aspects of vision, such as shape and form, color, contrast, distance and depth, and movement or motion. This parallel processing seems to happen independently in different patches of visual cortex. The results are recombined as these cortical areas communicate with other parts of the cerebral cortex – notably parts of the temporal lobe – plus the language centers and other areas. By such interactions we become aware of the color, shape, motion, distance, identity, name, and meaning of what we see.

At the optic chiasma, fibers from the left of each retina (green) join, as do those from the right (red). The fibers pass to the lateral geniculate nuclei (LGNs). Each LGN sends signals along optic radiations to its primary visual cortex. Side branches of the optic tracts (orange) feed data about vision to the pretecta and the Edinger Westphal nuclei, and back via ciliary ganglia to the constrictor pupillae muscles, which control pupil size. Other side branches (blue) go to the superior colliculi which, with the LGNs, direct the gaze to anything unusual (the visual startle reflex). These regions are also involved in the visual tracking reflex which allows moving objects to be followed.

The cat has no color and no apparent depth or movement. Yet it is instantly recognizable as a cat. In the visual cortex, nerve signals from the eyes are sorted into different aspects of vision: line, shape, color, movement, and distance or depth. Yet from this isolated feature of an image, simply a black shape, signals may stimulate neurons "higher" in the visual hierarchy, which code for familiar shapes such as a face, cat, dog, or car.

Stereoscopic, or binocular, vision allows us to judge distance and see objects, such as a cube, in three dimensions. Because of the half crossover at the optic chiasma, the left visual cortex receives signals from the left side of both retinas. In each visual cortex, signals can be directly compared to detect differences in angle, shading, and perspective that result from the two eyes looking at a three-dimensional object from slightly different viewpoints.

See also

INPUTS AND OUTPUTS
▶ Body links 80/81

▶ Seeing the light 88/89

▶ Contrasting views 90/91

▶ Active vision 94/95

▶ The big picture 114/115

▶ Survival sense 122/123

SURVEYING THE MIND
▶ Discovering through damage 24/25

▶ Brain maps 30/37

FAR HORIZONS
▶ The infinite store? 130/131

▶ Parallel minds 138/139

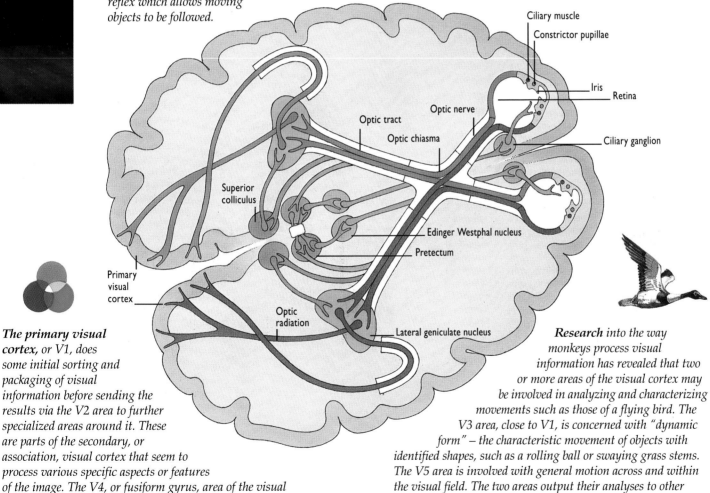

Ciliary muscle

Constrictor pupillae

Iris

Retina

Optic nerve

Optic tract

Optic chiasma

Ciliary ganglion

Superior colliculus

Edinger Westphal nucleus

Pretectum

Primary visual cortex

Optic radiation

Lateral geniculate nucleus

The primary visual cortex, or V1, does some initial sorting and packaging of visual information before sending the results via the V2 area to further specialized areas around it. These are parts of the secondary, or association, visual cortex that seem to process various specific aspects or features of the image. The V4, or fusiform gyrus, area of the visual cortex, for example, is apparently concerned mainly with the analysis and comparison of colors and contrasts.

Research into the way monkeys process visual information has revealed that two or more areas of the visual cortex may be involved in analyzing and characterizing movements such as those of a flying bird. The V3 area, close to V1, is concerned with "dynamic form" – the characteristic movement of objects with identified shapes, such as a rolling ball or swaying grass stems. The V5 area is involved with general motion across and within the visual field. The two areas output their analyses to other areas, where the features of the scene are reintegrated into a complete view of the world.

Active vision

Our visual perception is based only partly on external reality – the brain makes up the rest as it goes along.

From the moment we emerge into light at birth, vision dominates our conscious perception. We learn with our eyes, and we learn to see. The brain is not merely a passive recipient of nerve signal patterns sent from the eyes. It learns to make endless assumptions, short cuts, and extrapolations, so the apparently seamless scenes we see are partly guesswork. For example, the lens of the eye focuses an image onto the retina which is upside-down and back to front. But we never know any different. A baby gradually realizes that an object's image on a certain patch of the retina corresponds to a certain position in front of the body, from which the object can be picked up. By multitudes of such correlations, we learn to link retinal images with the position of an object in the physical world.

There are no rods and cones where the axons of millions of ganglion cells converge in the retina to form the optic nerve. This is the blind spot, insensitive to light – a "black hole" in the visual field. But the brain learns to extend lines, shapes, and colors into the blind spot. It also borrows input from the other eye, which usually looks from a slightly different angle, so the two blind spots rarely coincide. Because blood vessels fan out on the inside surface of the retina, light rays cast spidery shadows on it, throwing millions of rods and cones into darkness wherever we look. Again, the brain's visual perception fills in from experience. But the brain's leaps of assumption can be hijacked by the artificial situations of visual illusions – not fooling the eye, but tricking the brain.

There is only one triangle in this Kanizsa illusion, but gaps in the lines and black circles imply that there is another – upside-down. When it detects unexpected holes and gaps, the brain searches for an explanation. Specific neurons carry out these tasks, using experience of common line patterns and shapes.

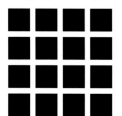

Dark patches lurk between the corners of these black squares – unless you stare at one, and then it disappears. With so many high-contrast shapes in such close proximity, the brain adds "common sense" connecting dark areas in peripheral vision. But if the image of a corner gap falls directly on the retina's cone-rich fovea (yellow spot), we see there is no such dark area.

Flicker fusion as in a movie results from the recovery of rods and cones after receiving light energy and firing nerve signals. The eyes are presented with a series of instant stationary snapshots, but the brain fills the gaps and perceives movement. In bright light, images fuse into a continuous scene when 60 per second are shown (a fly sees them as still and separate up to 300). In the dim light of a cinema, this flicker fusion frequency drops to 10 or below. So 24 images or movie frames per second produce an illusion of smooth movement.

Is it an elegant young lady in a hat looking over her shoulder, or an old crone resting her pointed chin on her chest? Once you perceive these two images, you can flip between them – but you cannot combine them. This may be due to two sets of neurons high in the visual processing system, each with a so-called projective field. One set corresponds to perception of each image, and they will be active alternately.

See also

INPUTS AND OUTPUTS
▶ Seeing the light
88/89

▶ Contrasting views
90/91

▶ Levels of seeing
92/93

▶ The big picture
114/115

▶ Survival sense
122/123

SURVEYING THE MIND
▶ Inside the mind
20/21

▶ Brain maps
30/37

FAR HORIZONS
▶ Active memory
132/133

▶ Parallel minds
138/139

▶ First thoughts
142/143

STATES OF MIND
▶ Emotional states
154/155

▶ The mind adrift
170/171

A BAD MEMORY FOR FACES

☤ A woman went to hospital to visit her husband who was recovering from a stroke, but when she entered the ward he showed no recognition. Only when she spoke did he realize that it was his wife. Prosopagnosia is the inability consciously to recognize or name familiar faces, even though the affected person knows that it is "a face."

Nerve signals produced in retinal ganglion cells, using information from light-sensitive rods and cones, go to various neurons at increasingly higher functional levels in the visual cortex. These neurons activate singly or in small groups only when a known whole pattern, such as a person's face, is received from their thousands of lower level inputs. This activation correlates with recognition and conscious awareness of a face. It may happen because each high-order neuron has its unique "projective field." This field consists of thousands of synaptic connections to neurons in other brain parts that code for related concepts such as the identity of the face, the person's name, associated memories such as the sound of the voice, and recent conversations, plus motor actions such as saying the name.

A stroke or brain injury may destroy some of the neurons or sever their synaptic links. If certain neurons in higher-level parts of the projective field are damaged, the face's identity, associations, and "meaning" are lost, even though lower-level neurons still register the image as a human face.

Is this a scene from **Gulliver's Travels?**
No, it is an Ames Room, which has distorted dimensions to fool our perception of perspective. Normally when we look at parallel lines, they converge with distance. This apparently happens here with the wall edges and floor tiles, so we perceive distance. And from the context of the room, we assume comparative sizes. The brain is fooled by tricks of perspective and its own expectations into thinking (wrongly) that the person on the left is the same distance away as the people on the right, and that the room is rectangular.

Architects sometimes play similar tricks with perspective to make it seem that buildings are taller, shorter, wider, or narrower than they really are.

One of the brain's most astonishing capacities is its ability to create its own images – dreams – without any visual input from the outside world. Dreams are usually visually coherent – seamless and with places and people as they would appear in normal life, which gives them their reality. It is thought that several visual areas in the brain are active during dreaming and that data from them are integrated before being fed back into the cortex as if they were coming from outside.

All of the colored images in this book, including the lips, are made from tiny dots in the three main printing-ink colors of yellow, cyan, and magenta, plus black. Under high magnification, the eye can distinguish these separate dots and their individual colors. At normal reading distance, however, the images that the dots throw onto the retina are so small and fine that the eye's light sensitive cone cells, which give us color vision, cannot discern them, so the dots merge into smooth areas of graduated color. If the printing quality is poor and the printing plates carrying the ink of the separate colors do not print on top of one another, dots in one of the individual colors can be seen as a "halo" of pure color around the image.

On a television screen the colors are produced by the stimulation of tiny fluorescent dots in the colors red, blue, and green. When the image is viewed from close up, the individual dots can be distinguished.

Children play for hours in a fantasy world based on props that give cues by sight. Such is the power of visual perception that a hat can transform reality and change behavior. So the child might think he is a cowboy, act like one, and see the world from a cowboy's viewpoint. Adults do the same – in theater. The associations generated by a visual image, as with other sensory inputs, are almost limitless since they come from the brain which has an almost infinite memory capacity.

Sensing sound

A sound that hits the ear takes a split second to register in the mind, but the journey is long and complex.

Ripples of alternating high and low pressure spreading through the air – sound waves – are produced by vibrating objects, like a bell or loudspeaker. The pitch of a sound is due to the number of vibrations per second and is measured in Hertz (Hz), with high frequency sounds being shrill and low ones being deep. You notice them because you have two tiny thin, flexible, skinlike membranes, each about the size of your little fingernail – your eardrums (tympanic membranes) – which vibrate in sympathy with sound waves. The vibrations are transferred to three tiny bones, the auditory ossicles – the malleus, incus, and stapes – and from them to the oval window, another thin membrane which is part of the wall of the fluid-filled cochlea. They continue as ripples of high and low pressure spreading through this fluid and shake a strip of membranes known as the organ of Corti.

The organ of Corti is a transducer, a device that can change energy from one form to another, in this case pressure waves in the cochlear fluid into patterns of nerve signals, which go to the cochlear nerve. So sound is transformed several times: from pressure waves in the air into vibrations in the solids of the eardrums and bones, to pressure ripples in the cochlear fluid, to motion in the organ of Corti, to electrical nerve signals which finally reach the hearing center in the brain. Here they are decoded, analyzed, compared with patterns in the memory, identified, and brought into your awareness.

When glass shatters, even if you are concentrating on something else, you instantly turn your head to look at the source of the sound. In this auditory reflex, signals are sent from part of the midbrain down to a part of the brain stem that has links to nerves which control muscles in the neck, and your head turns.

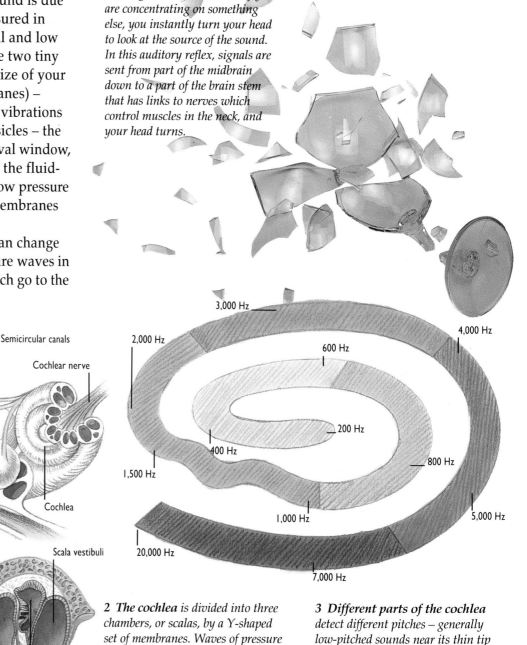

1 Sound waves funneled into the ear canal bounce off the eardrum, making it vibrate. The vibrations pass via the incus, malleus, and stapes to the oval window and into the cochlea.

2 The cochlea is divided into three chambers, or scalas, by a Y-shaped set of membranes. Waves of pressure change make the membranes vibrate.

3 Different parts of the cochlea detect different pitches – generally low-pitched sounds near its thin tip and shrill sounds at its wider base.

Semicircular canals

Cochlear nerve

Incus

Malleus

Eardrum

Ear canal

Stapes

Cochlea

Scala vestibuli

Scala tympani   Scala media

3,000 Hz
4,000 Hz
2,000 Hz
600 Hz
200 Hz
400 Hz
1,500 Hz
800 Hz
1,000 Hz
5,000 Hz
20,000 Hz
7,000 Hz

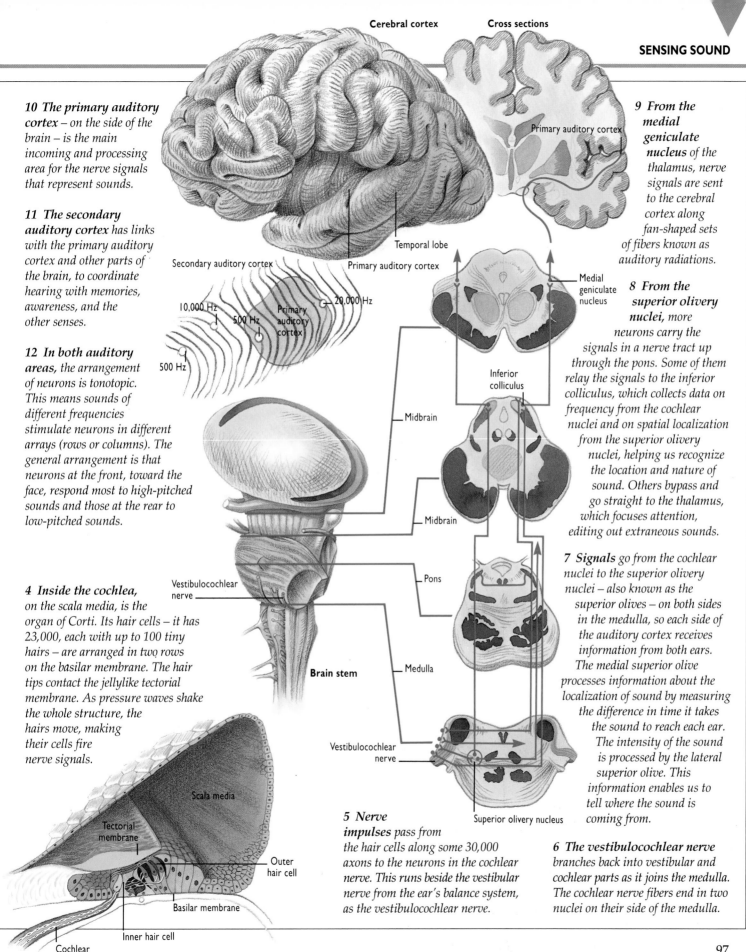

Cerebral cortex

Cross sections

Primary auditory cortex

Temporal lobe

Secondary auditory cortex

Primary auditory cortex

10,000 Hz

20,000 Hz

Primary auditory cortex

500 Hz

500 Hz

Medial geniculate nucleus

Inferior colliculus

Midbrain

Midbrain

Pons

Vestibulocochlear nerve

Medulla

Brain stem

Vestibulocochlear nerve

Superior olivery nucleus

Scala media

Tectorial membrane

Outer hair cell

Basilar membrane

Inner hair cell

Cochlear nerve

10 The primary auditory cortex – on the side of the brain – is the main incoming and processing area for the nerve signals that represent sounds.

11 The secondary auditory cortex has links with the primary auditory cortex and other parts of the brain, to coordinate hearing with memories, awareness, and the other senses.

12 In both auditory areas, the arrangement of neurons is tonotopic. This means sounds of different frequencies stimulate neurons in different arrays (rows or columns). The general arrangement is that neurons at the front, toward the face, respond most to high-pitched sounds and those at the rear to low-pitched sounds.

4 Inside the cochlea, on the scala media, is the organ of Corti. Its hair cells – it has 23,000, each with up to 100 tiny hairs – are arranged in two rows on the basilar membrane. The hair tips contact the jellylike tectorial membrane. As pressure waves shake the whole structure, the hairs move, making their cells fire nerve signals.

5 Nerve impulses pass from the hair cells along some 30,000 axons to the neurons in the cochlear nerve. This runs beside the vestibular nerve from the ear's balance system, as the vestibulocochlear nerve.

9 From the medial geniculate nucleus of the thalamus, nerve signals are sent to the cerebral cortex along fan-shaped sets of fibers known as auditory radiations.

8 From the superior olivery nuclei, more neurons carry the signals in a nerve tract up through the pons. Some of them relay the signals to the inferior colliculus, which collects data on frequency from the cochlear nuclei and on spatial localization from the superior olivery nuclei, helping us recognize the location and nature of sound. Others bypass and go straight to the thalamus, which focuses attention, editing out extraneous sounds.

7 Signals go from the cochlear nuclei to the superior olivery nuclei – also known as the superior olives – on both sides in the medulla, so each side of the auditory cortex receives information from both ears. The medial superior olive processes information about the localization of sound by measuring the difference in time it takes the sound to reach each ear. The intensity of the sound is processed by the lateral superior olive. This information enables us to tell where the sound is coming from.

6 The vestibulocochlear nerve branches back into vestibular and cochlear parts as it joins the medulla. The cochlear nerve fibers end in two nuclei on their side of the medulla.

See also

INPUTS AND OUTPUTS
▶ Body links 80/81

▶ The mind's ear 98/99

▶ The big picture 114/115

▶ Survival sense 122/123

SURVEYING THE MIND
▶ Comparing brains 18/19

▶ Brain maps 30/37

FAR HORIZONS
▶ Word power 136/137

Parallel minds 138/139

▶

STATES OF MIND
The mind adrift 170/171

The mind's ear

Hearing is a long way from being a passive sense – the mind can do much to manipulate what the ears detect.

First the violins lead the orchestral swell, then come the heavy-rock guitar power chords. As you listen intently to a favorite piece of music, you might concentrate on and pick out the bass line or the noise of the cymbals. This is similar to the way that you direct your eyes and stare intently at an object that interests you visually. But this auditory "tuning in" to listen to a particular instrument is different to visual tuning in. Your ears do not move, and thus they transduce all the sound waves they detect and send all the resulting nerve signals along the auditory nerves, so you must tune in within your brain, by active auditory perception. You concentrate your conscious awareness on a certain instrument by picking out and following its characteristic range of fundamental frequencies and harmonic overtones.

"Music" consists of sounds whose frequencies or pitches have mathematical relationships, and which are pleasing and harmonious to our auditory perception. "Noise" is a mixture of unconnected, discordant sounds. As you listen to your chosen musical piece, you may be distracted by a sharp noise from the side, and you try to localize it, or gauge its direction.

This is carried out partly by the two pairs of superior olivary nuclei in the brain's medulla. Because the speed of sound is relatively slow, about 1,115 feet/sec (340 m/sec), sound waves from the side arrive at the nearer ear slightly earlier than at the farther ear. This time difference is less than one-thousandth of a second, yet it is detected by neurons in the medial superior olivary nuclei. To enhance it, you may tilt your head and cock an ear in the supposed direction of the sound.

The sounds are also slightly louder or more intense in the nearer ear compared to the farther one. This volume difference is detected by neurons in the lateral superior olivary nuclei. The two pairs of nuclei send summary signals to the midbrain's inferior colliculus, for relaying to the superior colliculus. This coordinates body reflexes and reactions, such as head-turning and eye-swiveling, in response to sound and sight inputs. As you turn your head, the time and intensity differences lessen, until the brain can no longer discern them – and you are facing the source of the noise.

Although this mechanism seems extraordinarily responsive, human ears have their limits, responding only to frequencies of 20–20,000 Hertz (Hz) and to volumes above 10 decibels (dB) – we simply cannot hear higher, lower, and quieter sounds. So, for example, if a dog pricks up its ears and looks in a certain direction when its owner has noticed nothing out of the ordinary, it is because it can hear ultrasonic sounds above the range of human hearing and as high as 60,000 Hz.

WHALE SONG

The seas are oceans of sounds. Sound waves travel faster through water, at 5,000 feet/sec (1,500 m/sec), than through air, and they fade less quickly, too. Many water animals, from sea snails to squid, fish, dolphins, and whales, use a huge array of sounds for numerous reasons.

Dolphins and other toothed whales navigate and find their prey in murky waters by echolocation or sonar (sound radar), like bats in air. They send out pulses of sound, some incredibly high pitched

A capella ("in the chapel") is group singing without accompaniment. It relies heavily on vocal harmonies. A harmony is a series of notes sung or played together which are combined according to certain musical rules and which sound suited and pleasing to hear. In fact, due to the physics of sounds and vibration rates, musical harmonies are based on simple mathematical relationships. For example, the note of middle C (c') has a pitch or frequency of 256 Hz. The "same note but higher" is upper C (c''), an octave up the musical scale. Its frequency is 512 Hz, twice that of middle C. The two notes blend virtually as one.

See also

INPUTS AND OUTPUTS
▶ Sensing sound
96/97

▶ Staying upright
112/113

▶ The big picture
114/115

▶ Survival sense
122/123

SURVEYING THE MIND
▶ Comparing brains
18/19

▶ Brain maps
30/37

FAR HORIZONS
▶ Meeting of minds
134/135

▶ Word power
136/137

▶ Parallel minds
138/139

STATES OF MIND
▶ The mind adrift
170/171

THE MAESTRO'S SILENT SUFFERING

One of music's greatest masters, Ludwig van Beethoven (1770–1827), noticed his hearing was failing when he reached the age of 30. Some 19 years later, his world fell completely silent, and he was forced to use conversation books in which his friends wrote questions to which he was able to speak a reply. Once a virtuoso pianist, he was also forced to give up performing, because he could no longer hear what he was playing. Yet he continued to compose, and wrote such masterpieces as the world-famous Ninth (Choral) Symphony, completed in 1823, when profoundly deaf.

From about 1800, when his ears began to buzz and whistle and high notes became inaudible, the great composer was thrown into personal turmoil. Yet from the anguish and suffering, as he lost what he termed his "noblest faculty," came some of his most emotionally charged works. It seems that the neural circuits which represent memory traces in the brain were unaffected, since Beethoven was able to compose and arrange entirely in his mind, even when he could not hear anything at all. However, it greatly affected his social world and his conversations and relationships with others. To hide the problem, he pretended to be absent-minded. To attempt a cure, he flew into violent tantrums, poured strange ointments into his ears, and tried many other remedies.

Historians and doctors since have discussed the possible cause of Beethoven's deafness, which may have been typhus or a similar infection. Another suggestion is Paget's disease, when the body's bone maintenance is disturbed and some bones, such as the skull, grow abnormally. This produces a characteristically large head and brow, seen in later portraits of the composer. It may also crush the auditory nerves leading from ear to brain – hence loss of hearing, but retention of musical memories.

(up to 250,000 Hz), and detect and analyze the echoes that reflect from objects. Sperm whales can produce such powerful bursts of sound that the energy vibrations in the water stun their prey.

Many great or baleen whales produce complex songs which seem to us like clicks, squeals, howls, grunts, and moans. Some of their frequencies extend into infrasound, 20 Hz and below, which is too low-pitched for us to hear. So whale songs are often speeded up by tape recording or electronically pitch-raised for our listening. In a 20-hour session, a young adult humpback whale (**far left**) may repeatedly sing more than 10 songs, each up to 20 minutes long and consisting of up to 10 repeated themes. Each group of whales has its own "dialect," although the songs are unique to each individual and evolve over weeks and also from year to year. They are sung mainly in the breeding season, and they may serve to attract females for mating or to warn invaders from their territory.

Sense of touch

An object may feel warm or cold, wet or dry, rough or smooth, soft or hard – and all are perceived by touch.

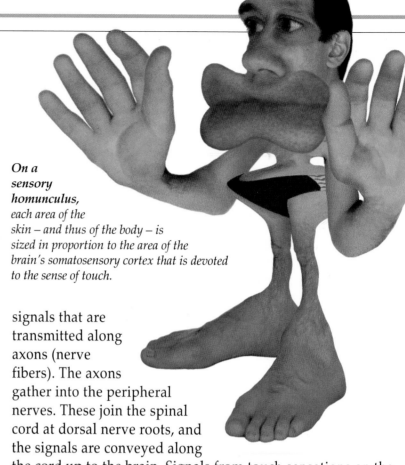

The detection of physical contact with the body – touch – might seem straightforward. But the fact that we can not only perceive but also tell the difference between a gentle stroke of the skin and a rough pinch suggests there is more to it. In fact, the word "touch" describes only some of the sensations such as pressure, temperature, movement, vibration, and pain – all based in the skin. Touch, together with proprioceptive sensations from within the body about the position and posture of various muscles, tendons, and joints, makes up the somatosensory system.

Skin sensors, or cutaneous exteroceptors, are microscopic structures embedded mainly in the dermis, the lower and living layer of skin just beneath the tough, dead outer epidermis. There are about six main kinds and their distribution varies over the body, from thousands per square millimeter in highly sensitive areas such as the lips and fingertips to fewer than a hundred per square millimeter in less sensitive areas such as the small of the back. When stimulated by mechanical distortion or thermal change, these sensors produce nerve

On a sensory homunculus, each area of the skin – and thus of the body – is sized in proportion to the area of the brain's somatosensory cortex that is devoted to the sense of touch.

signals that are transmitted along axons (nerve fibers). The axons gather into the peripheral nerves. These join the spinal cord at dorsal nerve roots, and the signals are conveyed along the cord up to the brain. Signals from touch sensations on the head and face are carried directly to the brain by the sensory branches of the trigeminal nerves (cranial nerves V).

In the brain, information about touch arrives at the somatosensory cortex, a strip across the top of each hemisphere, just behind the motor cortex. Here it is analyzed and, after further processing in the brain's association areas, details about the type of touch enter our conscious awareness.

The brain has a "touch-map" of the body *on its somatosensory cortex, or touch center, a strip around and down the side of each parietal lobe. Different-sized patches of the center are devoted to certain areas of skin, according to their degree of sensitivity. For example, the thumb has as much cortex devoted to it as the whole of the leg.*

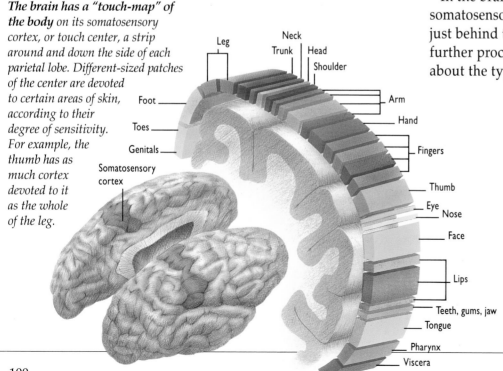

Leg
Neck
Trunk
Head
Shoulder
Foot
Arm
Toes
Hand
Genitals
Fingers
Somatosensory cortex
Thumb
Eye
Nose
Face
Lips
Teeth, gums, jaw
Tongue
Pharynx
Viscera

TOUCH TYPES

Walk on a feather, and you notice little. Step in a puddle, and it feels wet and cold. Stand on a pin, and you feel a painful stab. These sensations show the many different aspects of touch. But how do we discern them?

Various types of sensors in the skin, cutaneous exteroceptors, detect touch. The largest and deepest are onion-shaped Pacinian endings, some over 1/25 inch (1 mm) long. They pick up heavy

READING BY TOUCH

Blind from the age of three, the teenager Louis Braille (1809–52) attended a demonstration at the National Institute for Blind Youth in Paris of a code of dots and shapes embossed on card, used by the French military for secret, silent communications. He immediately realized its potential and began to adapt the complex code for normal reading purposes. By the age of 15, he had introduced his dot-based printing-reading system for visually handicapped people, which he continued to improve over the years.

The Braille system, which was standardized in 1932, uses cells, each being a pattern of up to six raised dots in three rows of two. Each cell (there are up to 63 combinations) represents a letter of the alphabet, a number, a punctuation mark, a common word such as "and" or "with," or a speech sound such as "ch." Usually one hand feels the dots with the sensitive fingertip skin, while the other hand feels for further lines.

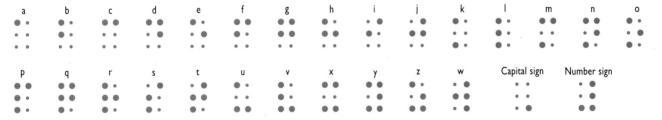

a   b   c   d   e   f   g   h   i   j   k   l   m   n   o

p   q   r   s   t   u   v   x   y   z   w   Capital sign   Number sign

See also

INPUTS AND OUTPUTS
▶ Body links
80/81

▶ The long junction
82/83

▶ On the move
84/85

▶ Guided motion
86/87

▶ Pain pathways
102/103

▶ Feeling pain
104/105

▶ Staying upright
112/113

▶ The big picture
114/115

SURVEYING THE MIND
▶ Brain maps
30/37

FAR HORIZONS
▶ Word power
136/137

STATES OF MIND
▶ Individual minds
184/185

pressure and fast vibrations, like those from a tuning fork. Smaller, egg-shaped Meissner's endings also detect vibrations, plus light touch. Both are fast-change mechanoreceptors, responding to brief mechanical stimuli by firing nerve impulses at an increased rate – but only while the stimulus is altering. Slow-change mechanoreceptors respond to more gradual alterations and continue to fire even under unchanging pressure. They include bulb-shaped Krause and sausage-shaped Ruffini endings. Merkel endings, which may project into the lower epidermis, pick up fast and slow mechanical changes and light touch. Free nerve endings are the most numerous microsensors. They detect most types of stimuli and are thermoreceptive and nociceptive – they feel heat, cold, and pain. Many different sensors may be stimulated at once, so it seems that our ability to distinguish various sensations lies – in part – in the brain's recognition of the pattern of sensory signals.

Pain pathways

Detection and perception of pain are essential if we are to protect ourselves from harmful injury.

The process of pain sensation begins with specialized, bush-shaped microsensors. They are known as free nerve endings, since the receptive parts of their membranes lack the characteristic structures of other microsensors associated with general touch. Free nerve endings are embedded in the skin at the junction between its outer epidermis and deeper dermis, and they also occur in many internal body parts. Different subpopulations of free nerve endings can detect several kinds of stimuli, including those that are mechanical (touch, pressure, and movement) and thermal (heat or lack of it). They are also the body's nociceptors, or "injury receivers," responding to potentially harmful events affecting body tissues.

Physical, chemical, microbial, or thermal injury causes tissue cells to rupture and spill out their contents, which include potassium and other ions. To limit and mend such damage, mast cells release histamine as part of the body's inflammatory response, which causes redness, soreness, fluid accumulation, and swelling. It seems that the chemicals released by these types of damage set off pulses of depolarization in the pain-specialized free nerve endings, suggesting that these receptors are mainly chemosensory (like smell and taste), detecting certain chemicals in their vicinity.

Nociceptive free nerve endings are distributed unevenly through the skin and body organs, which explains why injury to the facial skin is more painful than that to the skin at the back of the thigh, and why a damaged artery hurts more than an equivalently damaged vein. The only parts of the body lacking nociceptors are the intestines and the brain itself. This means that once the scalp, skull, and meninges (protective layers over the brain) are anesthetized, a surgeon can operate on a brain while the conscious patient feels little.

The nerve fibers carrying pain signals from the main body and limbs group into peripheral nerves that lead to the spinal cord. In the cord, pain signals are carried up to the brain along two pathways – the lateral spinothalamic tract and the spinoreticulothalamic tract. Signals concerning pain in the region of the face and head are conveyed directly to the brain by the cranial nerves.

Both touch and pain sensations are dealt with by the somatosensory system. But while touch signals are sent largely unmodified to the somatosensory cortex, for conscious perception, signals representing pain can be modified and even blocked by the activity of neurons in many parts of the spinal cord and in the brain itself.

YOU CAN RUB IT BETTER

An understanding of neural circuits in the spinal cord explains why rubbing the site of a sharp pain eases the hurt. Inputs from pain and touch neurons taking signals along their fibers to the spinal cord modulate the firing rate of interneurons in the spinal cord and in projection fibers, which carry signals toward the brain. The rate at which these signals are sent to the brain determines what is perceived.

Activity rate

	Low
	Medium
	High

With low input from both pain and touch fibers, the interneuron continues its normally high firing rate and inhibits the activity of the projection neuron so no pain is felt.

No pain

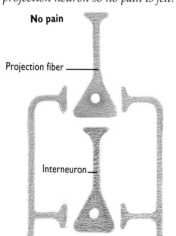

Projection fiber

Interneuron

When the pain fiber fires faster, it both inhibits the interneuron and excites the projection neuron. This causes the projection fibers to fire much faster, and so pain is perceived.

Pain only

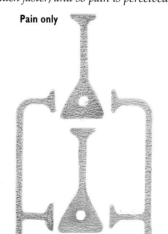

Rubbing activates the touch fiber, exciting the projection neuron and the interneuron, which resumes some inhibition of the projection fiber, and pain signals reduce.

Pain and touch

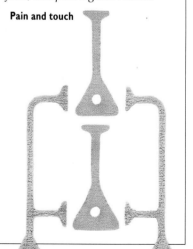

Pain fiber

Touch fiber

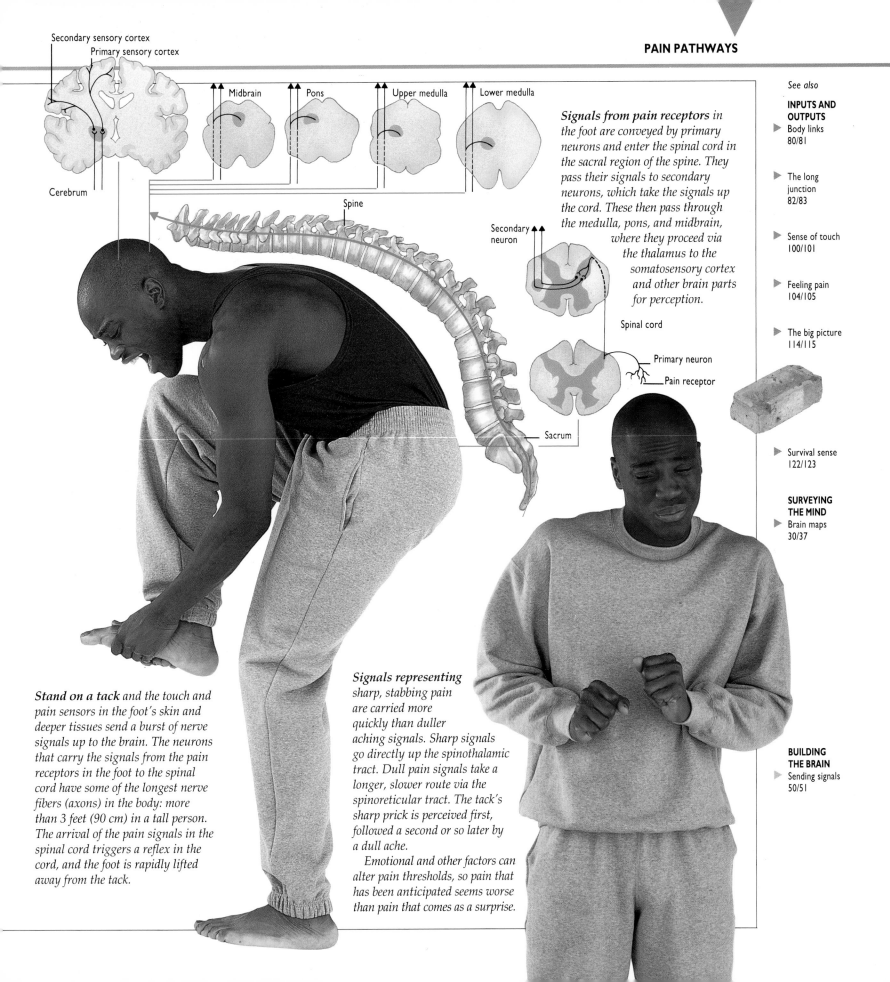

Secondary sensory cortex

Primary sensory cortex

Cerebrum

Midbrain

Pons

Upper medulla

Lower medulla

Spine

Secondary neuron

Spinal cord

Primary neuron

Pain receptor

Sacrum

Signals from pain receptors in the foot are conveyed by primary neurons and enter the spinal cord in the sacral region of the spine. They pass their signals to secondary neurons, which take the signals up the cord. These then pass through the medulla, pons, and midbrain, where they proceed via the thalamus to the somatosensory cortex and other brain parts for perception.

See also

INPUTS AND OUTPUTS
► Body links 80/81

► The long junction 82/83

► Sense of touch 100/101

► Feeling pain 104/105

► The big picture 114/115

► Survival sense 122/123

SURVEYING THE MIND
► Brain maps 30/37

BUILDING THE BRAIN
► Sending signals 50/51

Stand on a tack and the touch and pain sensors in the foot's skin and deeper tissues send a burst of nerve signals up to the brain. The neurons that carry the signals from the pain receptors in the foot to the spinal cord have some of the longest nerve fibers (axons) in the body: more than 3 feet (90 cm) in a tall person. The arrival of the pain signals in the spinal cord triggers a reflex in the cord, and the foot is rapidly lifted away from the tack.

Signals representing sharp, stabbing pain are carried more quickly than duller aching signals. Sharp signals go directly up the spinothalamic tract. Dull pain signals take a longer, slower route via the spinoreticular tract. The tack's sharp prick is perceived first, followed a second or so later by a dull ache.

Emotional and other factors can alter pain thresholds, so pain that has been anticipated seems worse than pain that comes as a surprise.

Feeling pain

How you perceive pain depends on a number of factors, from state of mind to time of day.

Most people agree on the type or nature of a certain pain – sharp, shooting, stabbing, dull, aching, sore, intermittent, episodic, and so on. But overlaid on this is each individual's state of body and mind. Perception of pain – its threshold, intensity, duration, and other factors – is heavily influenced by feelings, emotions, and knowledge. One reason for this is descending pain control. Nerve signals from the brain go along descending tracts in the spinal cord to modify the pain-related sensory inputs and their neural circuits within the cord.

An important part of the system involves the periaqueductal gray area (PAG) around the fluid-filled cerebral aqueduct between the third and fourth ventricles. Its neurons normally inhibit the activity of groups of neurons in the raphé nuclei, deep in the medulla. Researchers discovered how it operates indirectly. They found that when opiate-type substances such as the pain-relieving (analgesic) drug morphine are taken, they fit into inhibitory receptors on the PAG neurons. The activated inhibitory receptors dampen down the activity of their PAG neurons, releasing the raphé neurons

Endorphins are the body's own analgesics and are made mainly in the pituitary and hypothalamus. In molecular shape, they resemble opiate drugs such as morphine. Both fit into opiate-receptor sites on neurons to activate internal pain-relief systems. Enkephalins, such as methionine, are related substances made in the brain, adrenal glands, and other organs.

Morphine

Methionine

A runner experiences a natural high induced by the release of endorphins into the body as the race progresses. This helps the athlete to continue through the pain barrier. When the course is completed, the high – a raised physical and mental state – subsides, and the inner pain-suppressing systems die down. Now the realization of the pain breaks through, and agony follows the ecstasy.

from their inhibition. The latter then send signals down the descending tracts to stimulate interneurons in the cord, which block the input of pain signals from peripheral nerves.

At certain times the body makes and releases its own analgesic substances, known as endorphins (endogenous morphinelike substances). For example, when the body is under stress, through physical exertion, say, endorphins activate the pain control systems and act as natural painkillers.

See also

INPUTS AND OUTPUTS
▶ Body links 80/81

▶ The long junction 82/83

▶ Sense of touch 100/101

▶ Pain pathways 102/103

▶ The big picture 114/115

▶ Rhythms of the mind 120/121

SURVEYING THE MIND
▶ Brain maps 30/37

FAR HORIZONS
▶ Conditioning the mind 128/129

STATES OF MIND
▶ Feeling low 168/169

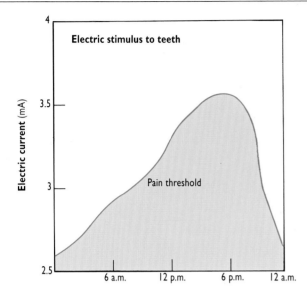

Electric stimulus to teeth

Electric current (mA)

Pain threshold

6 a.m.　　12 p.m.　　6 p.m.　　12 a.m.

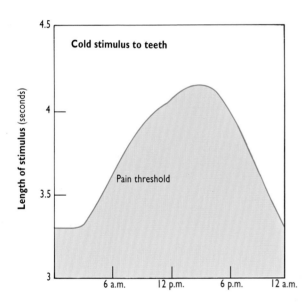

Cold stimulus to teeth

Length of stimulus (seconds)

Pain threshold

6 a.m.　　12 p.m.　　6 p.m.　　12 a.m.

How much pain you feel varies with time of day, so make an appointment with your dentist in the afternoon if you think you are especially sensitive to pain. Experiments have shown that in the average day, experience of pain waxes and wanes, with the lowest thresholds early and late. When irritation or discomfort turns to actual pain, you have reached your pain threshold. But it is a very individual and subjective experience, related to a host of general body variables. These include hormonal and other chemical levels (and even, in women, the phase of the menstrual cycle), the wake–sleep cycle and biorhythms, your level of hunger, and your emotional state, such as happy or depressed.

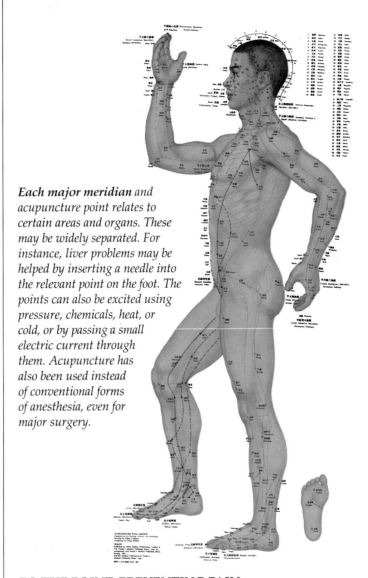

Each major meridian and acupuncture point relates to certain areas and organs. These may be widely separated. For instance, liver problems may be helped by inserting a needle into the relevant point on the foot. The points can also be excited using pressure, chemicals, heat, or cold, or by passing a small electric current through them. Acupuncture has also been used instead of conventional forms of anesthesia, even for major surgery.

TO THE POINT: PREVENTING PAIN

The ancient eastern tradition of acupuncture involves inserting fine needles into the body at specific points, to relieve pain, to reduce feeling and sensation, and to help treat illness and disease. Acupuncture points lie on channels known as meridians, along which chi, or "life force," energy circulates around and through the body. Imbalance in energy distribution along the meridians may lead to poor health, but acupuncture can restore the balanced flow and with it health. One possible explanation for acupuncture's effectiveness – in terms of conventional science – is that needle insertion stimulates peripheral nerves to relay signals to the central nervous system, triggering the release of endorphins for natural pain relief.

A matter of taste

Food without flavor would be like the beach without sunshine, so how do we taste what we eat?

Complaining that it tastes horrible, a sick child refuses to take some medicine. An adult sips it and says it just tastes bland. This scenario has its roots in sensory and neural fact. Taste (gustation), like smell (olfaction), is a chemosense – it detects the presence of certain chemicals. The individual "tasters" are chemosensory receptor cells, or chemosensors. They are shaped like the segments of an orange and are grouped with supporting cells into an orange-shaped cluster of 25 to 50 cells known as a taste bud.

The specialized chemosensors are short-lived, lasting only 10 days, but they are replaced within 12 hours. The average human adult mouth has up to 10,000 taste buds, mostly on the tongue, but also on the back of the upper mouth (palate) and down toward the throat. On the tongue, taste buds are grouped mainly on the sides and around the bases of the papillae, the small lumps and bumps on the tongue's upper surface, visible to the naked eye. Each taste bud is embedded in the covering layer, or epithelium, with a small hole that opens to the surface. Dissolved chemicals from food and drink seep through the hole to the chemosensors, whose tips have tufts of tiny hairs, or microvilli, that detect the chemicals.

Why do lovers of hot and spicy foods put up with that burning feeling when eating chili peppers? This sensation is not taste at all. Pain-type nerve endings in the tongue and mouth are stimulated by the chemical capsaicin, found in many peppers. The receptors send signals along the trigeminal nerve to the brain. Here, any sweet flavors already being tasted are heightened, and the release of endorphins – the body's natural painkillers – is triggered, causing feelings of wellbeing and pleasure. The result: chilis in a meal boost other flavors and make you feel good.

TASTY FEET

 In humans, taste and smell are separate sensory systems. Smell deals with airborne chemicals, taste with waterborne chemicals in foods and drinks. But in other animals, the distinction is more blurred. For example, a fish cannot detect airborne chemicals, only waterborne ones. So in the animal kingdom, biologists group taste, smell, and similar detecting systems together as chemosenses.

The distribution of the chemosensor cells on an organism is tailored by evolution to the lifestyle of the creature. Humans have them in the mouth, which is the first place we come into intimate contact with food. Some fish have them inside the mouth and nasal area, some on the outside of the snout and head, and others along the body. The catfish's body is covered with chemosensors, making it like a "living tongue." Flies, such as the housefly and blowfly (**right**), have chemosensors on their feet, which are especially sensitive to sweet, high-energy substances. So the fly knows at the moment of touchdown whether an object is suitable to eat.

See also

INPUTS AND OUTPUTS
▶ Body links
80/81

▶ On the scent
108/109

▶ The primal
sense
110/111

▶ The big picture
114/115

▶ Dealing with
drives
118/119

▶ Survival sense
122/123

SURVEYING THE MIND
▶ Comparing
brains
18/19

▶ Brain maps
30/37

FAR HORIZONS
▷ Conditioning
the mind
128/129

The many flavors of food *are said to be based on four main ones:*
sweet (strawberries, for instance), salty (chips), sour (lemon), and bitter (arugula).
These are sensed mainly on the tip, front sides, rear sides, and back of the
tongue, respectively, but attempts to link these basic flavors to types, groups,
or sites of chemosensors (taste cells) have failed. In addition, the tongue,
gums, lips, and mouth lining have somatosensory receptors, which respond
to touch, pressure, moisture, heat, cold, and other features of food. The brain
combines all this with the even more sophisticated input of the smell system.
What we think of as simply "taste" is a combination of touch, smell, and taste.

It seems that certain chemicals – the ones we taste – bind or
lock onto receptor sites on the chemosensors' cell membranes.
This probably opens sodium and other ion channels, causing
depolarization in the membrane, which may in turn generate
a nerve signal. The lower portions of the chemosensors have
synapses with the 20 to 30 nerve fibers supplying each taste
bud. Each cell probably has many different receptor sites,
which respond to a variety of tastes with a complex pattern
of signal-firing. When suitably stimulated by increased firing
rates from the chemosensors, the fibers conduct nerve signals
to the brain.

As a person ages, whole taste buds die, leaving possibly
fewer than 5,000 in old age. The rate of chemosensor cell
replacement also slows, so each individual taste bud has
fewer chemosensors. This is why in general older people have
less sensitive taste than younger people, hence the problem
with the foul-tasting medicine.

Taste signals *from the chemoreceptor cells on each side*
of the tongue and back of the mouth travel along three
pairs of nerves to the brain. From the front two-thirds of
the tongue, they go along a branch of the facial (VII cranial)
nerve; from the rear third, their route is via the lingual
branch of the glossopharyngeal (IX cranial) nerve; and
from the palate and upper throat, it is along the superior
laryngeal branch of the vagus (X cranial) nerve.

All the signals arrive in a region known as the nucleus
solitarius in the medulla. They then pass along more fibers
to the thalamus – the brain's relay station. This sends
signals to the primary and secondary gustatory areas or
"taste centers," near the somatosensory area or "touch
center" of the cerebral cortex. Nerve fibers also connect
the taste system to the hypothalamus (which controls
appetite) and the limbic system (which deals with emotions).
This is why taste can affect feelings of hunger and mood.

On the scent

A sniff of the air sets in motion a complex train of actions that can lead to the perception of a smell.

Like taste, our sense of smell, or olfaction, is a chemosense: it detects the presence of chemicals, in this case odorants, or odor molecules. These chemicals arrive in the nose floating on a stream of indrawn breath and land on the mucus-coated interior of the nasal cavity; if they dissolve in the watery mucus, they stand a chance of detection. At the top of the nasal cavity in each side of the nose is a thumbnail-sized patch, the olfactory epithelium. This has about 10 million olfactory receptor cells. The main dendrite of each cell extends downward to the nasal cavity. It has a swollen tip bearing up to 20 microscopic hairs, or cilia, that "float" in the nasal mucus. Individual olfactory receptor neurons live for 30 to 35 days; then they are replaced by cell division.

When specific odorants lock into receptor sites on the cilia, they can generate nerve impulses in an olfactory cell. These travel from the cell's body along its axon, which projects through the thin skull bone just above, and into the olfactory bulb. The diffuse webs – one for each nostril – of axons from olfactory cells passing through the skull make up the paired olfactory nerves, also known as the I (first) cranial nerve. The bulbs process the signals and relay the results to the rest of the brain, including some so-called primitive parts involved in functions such as emotion and memory. This "hard-wiring" into these regions, part of what is known as the paleocortex, is a major difference between smell and the other senses.

Sawdust

Banana

"Smell maps" can be derived from EEG readings of the wavelike electrical signals recorded across the olfactory bulb of a rabbit. The computer image shows the distribution of the amplitude, or strength, of the waves which represent nerve signal activity. The result is like a contour map, with the most active areas resembling hills.

A rabbit trained to recognize the odor of sawdust produces a specific pattern in its bulb (***above left***), derived from the bursts of waves produced as it inhales. When it is then familiarized with the smell of banana, a different pattern results (***above***). But the map can change with learning and experience.

From first sniff to recognition, an odor passes along many pathways: through the olfactory system – from nose to olfactory cortex – to limbic system, thalamus, and frontal cortex.

In the olfactory bulb, nerve impulses from olfactory cells enter one of hundreds of olfactory glomeruli – small ball-like tangles of axons, synapses, dendrites, and cell bodies. Next the signal goes along the olfactory tract to the secondary olfactory cortex. The anterior

olfactory nucleus links the bulbs from the two nostrils via the anterior commissure. The olfactory tubercle and the pyriform cortex project to other olfactory cortical regions and to the medial dorsal nucleus of the thalamus; they are involved in conscious perception of smell. The last two, with the amygdaloid complex and the entorhinal area, which in turn projects to the hippocampus, are pathways to the limbic system, which is why smells evoke memories and emotions.

THE WORLD IN SMELLS

🐾 A dog probably lives in a world where smells are as important to it as vision is to us. A scent-detection dog, for instance, has an area sensitive to smell 30 times larger than ours, containing 10 times as many olfactory receptor cells. Dogs also have a proportionally larger area of cortex devoted to analyzing smells. An average human can discern some smells at concentrations of less than 1 part in 20 billion, but scent-detection dogs can detect odors at least 10,000 times weaker.

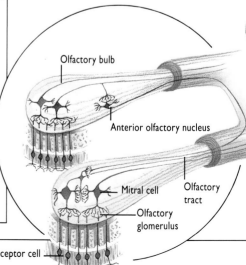

Olfactory bulb

Anterior olfactory nucleus

Mitral cell

Olfactory tract

Olfactory glomerulus

Olfactory receptor cell

See also

INPUTS AND OUTPUTS
▶ A matter of taste 106/107

▶ The primal sense 110/111

▶ The big picture 114/115

▶ Survival sense 122/123

SURVEYING THE MIND
▶ What is a brain? 12/13

▶ Comparing brains 18/19

▶ Probing the mind 26/27

▶ Brain maps 30/37

BUILDING THE BRAIN
▶ Support and protection 42/43

▶ The brain's cells 44/45

Sawdust

When the rabbit inhaled sawdust again after it had learned the banana smell, a different pattern emerged. This shows how such maps are not specific to the smell itself, but appear to change with time and according to experience. Furthermore, another rabbit will almost certainly have a different map for sawdust.

Acting on the signals from a region of about 25,000 olfactory receptor cells in the nose, each glomerulus in the olfactory bulb reacts to certain odorants. The number of receptors activated indicates the strength of the smell stimulus, and their position in the nose supplies information as to the nature of the scent.

Messages are relayed from one glomerulus to the next, probably by periglomerular cells, and a pattern of activity, as shown in the sawdust and banana "maps," is generated which carries information about the odor. For this information, which is in the form of a burst of nerve signals, to get through to the rest of the brain, it has to be strong enough to survive the inhibitory effects of the granule cells in the bulb. Successful messages are carried along the axons of mitral and tufted cells, which form the olfactory tract, into the olfactory cortex.

Here the signals have to pass through an intermediate layer of cells, the superficial pyramidal cells. They synapse with and excite stellate cells, as well as deep pyramidal cells. But when excited, the stellate cells inhibit the deep pyramidal cells, creating a loop of excitation (red) and inhibition (blue) that has the effect of generating bursts of nerve signals which are then transmitted to other brain regions.

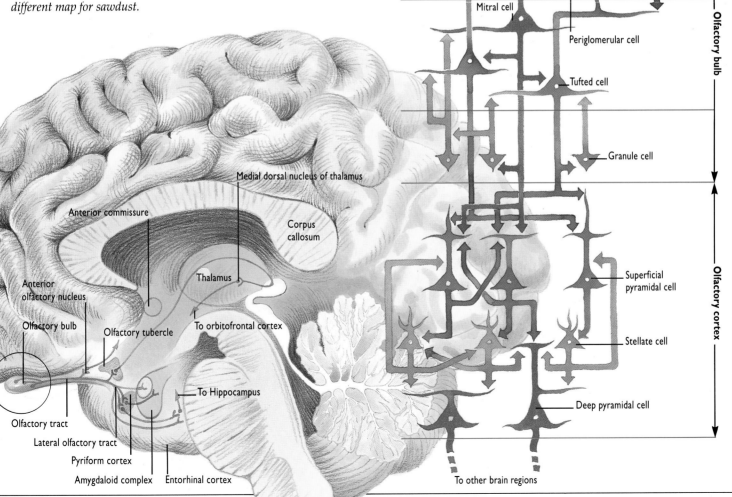

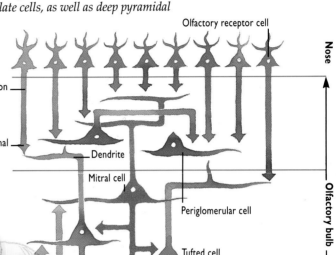

Olfactory receptor cell

Nose

Axon

Axon terminal

Dendrite

Mitral cell

Periglomerular cell

Tufted cell

Olfactory bulb

Granule cell

Medial dorsal nucleus of thalamus

Anterior commissure

Corpus callosum

Thalamus

Anterior olfactory nucleus

Superficial pyramidal cell

Olfactory bulb

Olfactory tubercle

To orbitofrontal cortex

Stellate cell

Olfactory cortex

To Hippocampus

Olfactory tract

Deep pyramidal cell

Lateral olfactory tract

Pyriform cortex

Amygdaloid complex   Entorhinal cortex

To other brain regions

The primal sense

Almost without our being aware of its presence, a scent can subtly affect the way we act and feel.

Unlike the other human senses, our sense of smell is linked directly by nerve pathways into the less sophisticated, more primitive, animal-like parts of the brain such as the limbic system and the sites involved in processing and storing memories. Nerve signals from the olfactory bulb, which is situated above the nasal cavity, can go directly to these parts without passing via the cortex, where they would stimulate our conscious awareness and allow us to think about them and modify our behavior accordingly. These direct pathways to primitive areas mean that certain smells have powerful effects on memories and emotions. We experience smells in a subconscious way, almost as the opposite of a hallucination. Whereas a smell is there, influencing our mind and behavior although we are not obviously aware of it, a hallucination is a conscious experience, but with no external cause or stimulus.

Since we experience smells in this manner, the way in which we respond to them may have something in common with the ways that animals react to them (and perhaps to other sensed stimuli). Our own awareness is dominated by "higher" human thought processes and by the senses of vision and sound. But many creatures experience the world primarily in terms of odors, and the use of smells is widespread in animal survival, communication, mating, and other types of behavior.

For instance, biologists have found that salmon migrate, sometimes over great distances, from the sea back to the stream where they were spawned, mainly by using their chemosenses to detect and follow the "scent-taste profile" of their home water. Pike leave their home area, where they lurk to ambush potential prey, in order to get rid of body wastes. This prevents the pike's excretory smell from building up in the home area and warning other fish of the hunter's presence. There are thousands of such examples in nature, right across the animal kingdom, ranging from worms and insects to various types of mammals. The chances are that we humans are more influenced by smells than we think.

A perfumier checks a blend in the never-ending search for new scents. Years of experience have given him a "trained nose." In fact, his nose is probably no more sensitive than most other people's, but he has trained his brain to focus on and isolate what it detects. Extracts of scents, smells, and essences are big business for many industries, from perfume and cosmetics to food and drink, soap and hygiene products, and air fresheners and detergents.

Peppermint

Floral

Ethereal

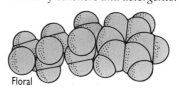

One theory of how odor molecules activate smell receptors in the nose is that they are shaped to fit into equivalently structured protein receptor sites. There may be several basic shapes, shown here. But smell seems to be based on pattern-recognition rather than being stimulus specific: most smells activate most of the olfactory cells, so the identity of the smell lies in the overall pattern of activation.

See also

INPUTS AND OUTPUTS
▶ Body links
80/81

▶ A matter
of taste
106/107

▶ On the scent
108/109

▶ The big picture
114/115

▶ Survival sense
122/123

SURVEYING THE MIND
▶ Comparing
brains
18/19

▶ Brain maps
30/37

FAR HORIZONS
▶ Conditioning
the mind
128/129

▶ The infinite
store?
130/131

▶ Active memory
132/133

▶ Male and female
148/149

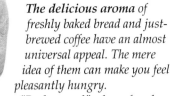

The delicious aroma of freshly baked bread and just-brewed coffee have an almost universal appeal. The mere idea of them can make you feel pleasantly hungry.

"Background" odors of such evocative foods are used in many ways by supermarkets and other organizations. The bakery or coffee shop may be loss leaders – as departments, they will not make any profit; indeed, they make a loss. But their wafting odors attract people, alter their mood, increase their sense of wellbeing, and perhaps – the store certainly hopes – make them spend more money. If there is no bakery or coffee shop, extracts of the relevant odors may be circulated in the air-conditioning system for the same purpose.

This is just one of many varied ways in which smells can affect our psyche subliminally. But the initial impetus is short-lived: we can usually only perceive a smell for about 30 seconds after it was first detected. This is due to a process known as habituation in which a sense stops perceiving what it detects once it has become accustomed to it. This happens more quickly in smell than in any of the other senses.

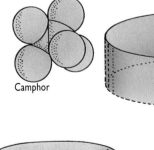

Musk

Camphor

CHEMICAL COMMUNICATION

Some animals release particular sorts of chemicals, known as pheromones, into the air or water or onto objects. These are chemosensed by fellow members of their species, whose behavior is altered as a result.

Humans are also thought to make use of pheromones, especially in the area of reproduction. These chemicals may be responsible for the phenomenon in which women who live or work together become synchronized in their menstrual cycles.

They may also explain why women in particular are often able to identify the sex of a person by the smell of his or her breath or underarm odor, and why some men can tell the stage of a woman's menstrual cycle from her vaginal odor.

Researchers are now striving to identify a pheromone that works as a sexual attractant between humans. It could have various uses and commercial rewards! But they have had little success so far.

Pheromones are used by animals for a variety of purposes. Female moths release a mating pheromone that can be detected by the feathery antennae of the males, up to 1 mile (2 km) away. The males follow the scent and mate with the females. Fish use pheromones to communicate, with individuals of the same or different species.

Staying upright

We take balance for granted, but the nervous system's "behind the scenes" action is highly sophisticated.

Balance, often called the sixth sense, is not really a sense at all. Rather, it is a process in which the brain, using nerve signals from a range of sensory inputs, works out which muscles – in the trunk, legs, arms, and so on – must move to keep an erect posture and equilibrium. One input is from the vestibular system inside the ears which detects head movements and gives information about head orientation in relation to gravity. Visual input helps work out the head's position, using verticals such as walls and trees, and horizontals such as floors.

A third input is from pressure sensors in the soles of the feet and other skin areas in contact with firm surfaces. These give data about the body's center of gravity and leaning angle. Fourth is proprioception – the body's inner positional-posture sense. Microscopic sensors in muscles, tendons, joints, and ligaments all over the body detect tension, elongation, and other changes. They send signals to the cerebellum and, via the spinal cord and thalamus, to the center for conscious positional sense in the cerebral cortex.

Coupled to these inputs are unconscious pathways that produce reflex reactions, many of which route through the cerebellum. For example, if you stand and rock back on your heels, you may automatically thrust your arms forward as a counterbalance in the extensor thrust reflex.

Eyes closed, arms by your side, and standing one-legged on a pillow, you are sure to totter as so many inputs to balance are blocked. There is no visual input; the arms cannot counterbalance; and, apart from being on only one foot, data from sole sensors is blurred due to the soft pillow. A wobbly toy's low center of gravity, by contrast, ensures a quick return to upright.

THE RIGHTING IS ON THE FALL

A cat's marvelous agility – especially the way it almost always lands on its feet – is not a result of feline magic or lives to spare. Cats are climbing animals, and over time, natural selection has determined that the cats that were best able to survive falls from a height passed on their ability to future generations.

When a cat falls, its finely tuned balance mechanism swings into action: first the head rotates back to horizontal, followed by the body, next the legs stretch out and the toes spread, ready for a soft landing. This righting reflex primarily involves structures in the inner ear: the utricle and saccule. Signals from them pass to processing centers (nuclei) in the upper medulla (the lower stalklike projection of the brain) and the cerebellum, for fast analysis and reflex-type reactions. The processes are similar, but slower and more ponderous, in humans.

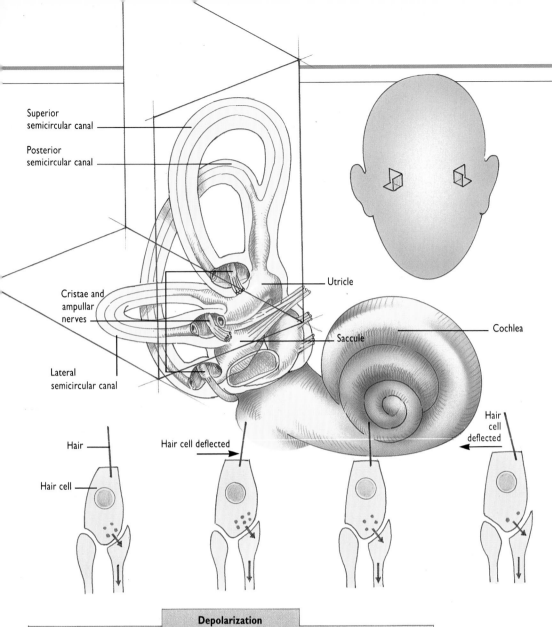

Superior
semicircular canal

Posterior
semicircular canal

Cristae and
ampullar
nerves

Lateral
semicircular canal

Utricle

Saccule

Cochlea

Hair
cell
deflected

Hair

Hair cell deflected

Hair
cell
deflected

Hair cell

Depolarization

Hyperpolarization

Nerve cell firing rate

The balance, or vestibular, organs of the inner ear are the three semicircular canals, the utricle, and the saccule. The semicircular canals sense chiefly acceleration, deceleration, rotation, and other motions of the head. They are almost at right angles to each other, occupying three planes of space, so whichever way the head moves, at least one is affected. The utricle and saccule, also known as the otolith organs, detect mainly the position of the head in relation to gravity's downward pull and also linear acceleration and deceleration.

The vestibular system's receptor is the hair cell. When a hair cell's hairs are bent one way, they convert their movement into depolarization of the cell, causing it to release more neurotransmitters which cross to the outgoing, or afferent, nerve and depolarize it. This raises its firing rate from its steady background rate.
  If the hairs are deflected the other way, they cause their hair cell to hyperpolarize, reducing its neurotransmitter release and thus hyperpolarizing the afferent nerve and reducing its firing rate. The change in the firing rate in the afferent nerves is used in the balance process.

See also
INPUTS AND OUTPUTS
▶ The long junction 82/83

▶ On the move 84/85

▶ Guided motion 86/87

▶ Sensing sound 96/97

▶ Sense of touch 100/101

▶ Survival sense 122/123

SURVEYING THE MIND
▶ What is a brain? 12/13

▶ Brain maps 30/37

BUILDING THE BRAIN
▶ Crossing the gap 52/53

▶ Unlocking the gate 58/59

▶ Discovering transmitters 60/61

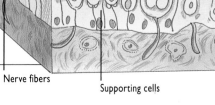

Hair cell

Hair

Calcium carbonate crystals

Nerve fibers

Supporting cells

In each utricle and saccule, the hairs of hundreds of hair cells are embedded in a jellylike membrane, containing calcium carbonate crystals. As the head moves or stays still, motion and gravity drag the membrane one way, then another, bending the hairs.

Hair cells in the ampulla of a semicircular canal cover the crista. Their tips are embedded in the jellylike cupulla. Head motion sets up currents in endolymph fluid in the crista which bend the cupulla and hairs. The nerve signals pass along the ampullar nerve.

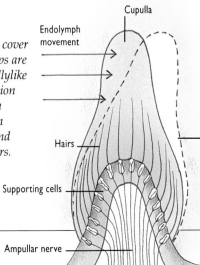

Cupulla

Endolymph movement

Hairs

Supporting cells

Ampullar nerve

Crista movement creates nerve impulses

The big picture

Engaged in a continuous exchange of signals with the cortex, the thalamus is the brain's sensory gatekeeper.

Millions of nerve signals, representing sensory information, pour into the brain every second. They concern not only outside events as detected by light, sound, smell, taste, and touch, but also internal variations such as change in posture and body temperature as well as levels of nutrients and hormones. But the brain can cope. Nerve tracts and connections route each batch of signals to appropriate sites in the brain for simultaneous analysis. A vital site is the thalamus, which consists of two egg-shaped structures about 1½ inches (4 cm) long in the core of the brain, below the corpus callosum and cerebral hemispheres.

The thalamus is a major center for switching, routing, and relaying sensory inputs to their specific areas of the cerebral cortex for analysis and processing. It has two-way relationships and integrating functions, receiving reciprocal signals from each of the sensory cortical areas and both sending and receiving signals to and from other more general "association" areas of cortex. In addition, the ventral parts of the thalamus have a role in motor responses. They receive signals from the cerebellum and basal ganglia, coordinate these with incoming sensory information, and send additional messages up to the premotor and motor cortex areas. Other thalamic functions include the general state of body awareness and arousal, linked to biorhythms and the wake–sleep cycle, and connections to the limbic system, associated with expressing feelings and emotions.

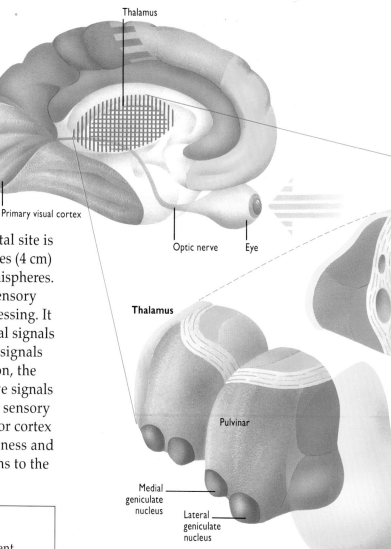

The thalamus, which nestles in the central region of the brain, has a core of gray matter, chiefly nerve cell (neuron) bodies and their connections – dendrites and synapses – covered with a thin layer of white matter, mainly nerve fibers, or message-carrying axons.

The gray matter in the thalamus is organized into "lumps," or nuclei, that are associated with certain inputs and outputs. In effect, part of the thalamus is a miniature version of the cerebral cortex. Each of its nuclear masses has two-way connections with a corresponding area of cerebral cortex, but the "map" of the thalamus has its sites shifted around slightly, compared to the map of the cortex.

Here, the regions of the thalamus that connect with regions of the cortex have been color-matched. The medial geniculate nucleus (above) has connections with the primary auditory cortex (right), and the lateral geniculate nucleus (above) links to the primary visual cortex (top and right).

THE COLOR OF WORDS

For as long as she could remember, a woman had seen colors when she heard words or letters. The colors were consistent, so she always saw yellow with hints of green when she heard the word "king," for instance.

The woman had synesthesia, the mingling or even swapping of sensory information in which stimulating one sense triggers conscious experience in another. In her case, words had colors, but all the senses can be confused. Thus spoken words have tastes or shapes, colors smell different, and touches on the skin have sound signatures. The condition occurs in about 1 in 25,000 people, mainly women. It seems to arise early in childhood and is consistent, involuntary, and non-suppressible: it "just happens." It rarely causes great suffering, indeed some synesthetics enjoy it.

One theory for the shared or mixed sensory experiences is that the sensory areas of the brain are "cross-wired" within the cortex or via the thalamus. We are born into this state, but – except in synesthetics – the developing conscious separates out the senses. Another idea is that it occurs in the limbic system – and that the limbic system is the main site where sensory information normally combines to form our "big picture" of conscious experience.

Sensory inputs from the eyes and some other sense organs are relayed through nuclei of the thalamus en route to the cerebral cortex. Visual nerve signals pass along the optic nerve and through the lateral geniculate nuclei on their way to the primary visual cortex for initial processing. Smell is the only sense that does not channel its nerve signals through the thalamus before they reach the cortex.

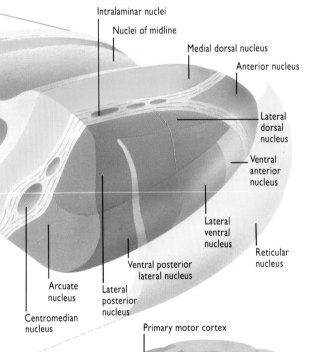

Intralaminar nuclei
Nuclei of midline
Medial dorsal nucleus
Anterior nucleus
Lateral dorsal nucleus
Ventral anterior nucleus
Lateral ventral nucleus
Reticular nucleus
Ventral posterior lateral nucleus
Lateral posterior nucleus
Arcuate nucleus
Lateral posterior nucleus
Centromedian nucleus

See also

INPUTS AND OUTPUTS
▶ Guided motion 86/87

▶ Levels of seeing 92/93

▶ Sensing sound 96/97

▶ Pain pathways 102/103

▶ A matter of taste 106/107

▶ On the scent 108/109

▶ Survival sense 122/123

A savage storm assaults the senses with sights, sounds, touch sensations, and movements. The sensory receptors feed their nerve messages, mainly via the thalamus, to the "thinking" cerebral cortex. But the cortex also sends signals back to the thalamus. Perhaps by such pathways inputs from some senses can be suppressed at the thalamic level, so that we can selectively attend to just one sensory input. We can choose to pay attention to the sight of the waves crashing against the rocks, or their sound, or the needles of sea spray splattering the skin and making it prickle. Meanwhile, should any part of the big picture demand attention – possibly because it poses some threat or because it becomes "interesting" for some reason – the system quickly directs awareness to that part.

SURVEYING THE MIND
▶ What is a brain? 12/13

▶ Brain maps 30/37

FAR HORIZONS
▶ The infinite store? 130/131

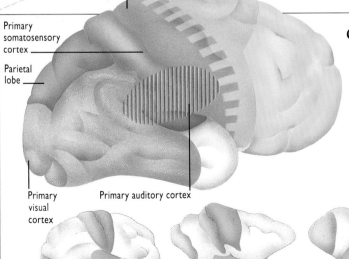

Primary motor cortex
Primary somatosensory cortex
Parietal lobe
Primary visual cortex
Primary auditory cortex

CLEVER ASSOCIATION

Relative to the size of the cortex, cerebral cortex areas dedicated to specific functions become smaller – in line with what is regarded as increasing intelligence – from creatures such as rats through to humans (see key left to comparisons). The non-dedicated cortex increases in proportion and may integrate processed sensory inputs to form a combined sensory experience, make decisions, and initiate actions – the "higher" processes of thought.

Motor
Somatosensory
Visual
Auditory
Olfactory
Association

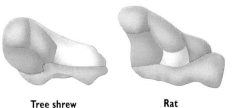

Human   **Chimpanzee**   **Tarsius**   **Tree shrew**   **Rat**

Keeping control

Not only is the brain the seat of reason, perception, memory, and emotions, but it also controls essential body processes.

The autonomic nervous system (ANS) controls and coordinates the body's "automatic" functions – the vital life processes such as heartbeat, blood pressure, digestion, excretion and, to an extent, breathing. We are usually unaware of ANS control because it is based in the brain stem, the "lower" part of the brain, where activities are mainly at the unconscious level. For example, in the medulla of the brain stem are two centers involved in heartbeat control: the cardio-acceleratory area (CAA) and, below it, the cardio-inhibitory area (CIA). Working together in a balanced way, they make up the cardioregulatory center, which adjusts the heartbeat rate according to sensory feedback from many parts of the body.

Other such complexes are the vasomotor center in the lower part of the pons and medulla, which regulates blood pressure, and the respiratory center, which is concerned with breathing. Many of these brain stem centers have links with the hypothalamus, the highest integrating area of the ANS. The hypothalamus is the main mediator between the autonomic, or unconcious, control systems in the brain stem and the voluntary, conscious "higher" mental activities in the cerebrum and other upper parts, especially the cerebral cortex. So if a threat is perceived, signals from the hypothalamus make the brain stem generate the appropriate physical responses.

In the kitchen, the chef checks the seasoning while the pressure regulator on the steam cooker keeps its internal conditions within set limits. Likewise the body uses neural receptors to test for chemicals, blood pressure, and many other internal conditions.
Chemosensors in the brain, blood vessels, and various organs detect levels of sodium, glucose, oxygen, carbon dioxide, and other substances in blood and body fluids. Barosensors in the walls of the main arteries, mainly in the neck and chest, respond to changes in blood pressure. Such sensors send their information to the brain stem's autonomic auto-control systems.

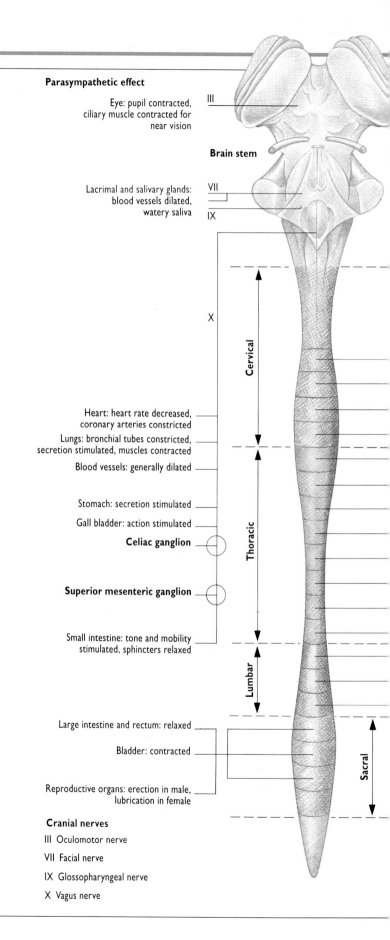

Parasympathetic effect

Eye: pupil contracted, ciliary muscle contracted for near vision — III

Brain stem

Lacrimal and salivary glands: blood vessels dilated, watery saliva — VII
IX

X

Cervical

Heart: heart rate decreased, coronary arteries constricted
Lungs: bronchial tubes constricted, secretion stimulated, muscles contracted
Blood vessels: generally dilated

Stomach: secretion stimulated
Gall bladder: action stimulated
Celiac ganglion

Thoracic

Superior mesenteric ganglion

Small intestine: tone and mobility stimulated, sphincters relaxed

Lumbar

Large intestine and rectum: relaxed

Bladder: contracted

Reproductive organs: erection in male, lubrication in female

Sacral

Cranial nerves
III Oculomotor nerve
VII Facial nerve
IX Glossopharyngeal nerve
X Vagus nerve

Sympathetic effect

Eye: pupil dilated, ciliary muscle relaxed for far vision

Salivary glands: blood vessels constricted, thick saliva

Superior cervical ganglion

Stellate ganglion

Heart: heart rate increased, coronary arteries dilated

Lungs: bronchial tubes dilated, muscles relaxed

Liver: glycogenolysis increased

Stomach: secretion inhibited

Gall bladder: action inhibited

Small intestine: tone and mobility inhibited, sphincters contracted

Celiac ganglion

Adrenal gland: epinephrine and norepinephrine secreted

Superior mesenteric ganglion

Skin: sweat glands stimulated, peripheral blood vessels constricted, hairs stand erect

Inferior mesenteric ganglion

Large intestine and rectum: contracted

Bladder: relaxed

Reproductive organs: orgasm

Paravertebral ganglia

The autonomic nervous system (ANS) has two major subdivisions, the sympathetic and the parasympathetic, which have balanced "push-pull" effects. Each exerts its control by sending nerve signals to many groups of muscles – those in arterial walls that control blood distribution and pressure, cardiac muscle in the heart, muscle layers in the walls of the intestines, glands and other internal organs, and skeletal muscles that move the body.

The parasympathetic subdivision's signals leave the brain stem via cranial nerves, in particular cranial nerve X (the vagus), and go to the chest and abdominal organs. They also go from the sacral portion of the spinal cord. The parasympathetic system takes the lead when the body is in its normal resting state and is relatively inactive. It promotes blood flow to the organs that deal with digestion and excretion, while keeping blood flow to the inactive skeletal muscles to a minimum; it also makes breathing easy and relaxed and the heartbeat steady.

The sympathetic subdivision sends its signals from the brain stem down the spinal cord and out along the spinal nerve roots of the cervical, thoracic, and some lumbar segments. The signals travel onward to the muscles via the necklacelike chains of lumps on each side of the backbone known as paravertebral ganglia. The sympathetic system takes over when the body is in a heightened state of awareness and is ready for action. It stimulates the heart, breathing, and skeletal muscles to work faster; tops off levels of high-energy glucose in the blood; and heightens conscious awareness and sensory perception, while temporarily shutting down the internal processes of digestion and excretion. In other words, it produces the classic fight-or-flight response, preparing the body for action.

CONTROL OF BREATHING

By focusing the mind on controlled breathing, practitioners of techniques such as yoga can achieve a state of deep relaxation. One possible reason for this effect is that the brain is "fooled" by this type of breathing. Slow and even breathing resembles that of sleep, and the brain sends out signals that are appropriate to this state. These then calm the body and mind.

But most of us rarely think about our breathing at all. Instead, several areas in the brain stem work together as the respiratory center to control it. They include the inspiratory and expiratory neural circuits in the medulla's respiratory rhythmicity area. These stimulate breathing in, then out, by "seesaw" alternation of activity every few seconds. Other areas are the apneustic and pneumotaxic areas in the pons. The former encourages strong inhalations with weak exhalations, to build up air volume in the lungs; the latter inhibits this activity. These areas are in turn influenced by data coming from other parts of the brain and body such as chemosensors that detect the blood's acidity (a guide to its carbon dioxide content) and barosensors, which monitor blood pressure. Rises in both of these stimulate faster, deeper breathing, as does information from the muscles and joints if they are active.

See *also*

INPUTS AND OUTPUTS
▶ Body links 80/81

▶ The long junction 82/83

▶ On the scent 108/109

▶ A matter of taste 106/107

▶ The big picture 114/115

▶ Dealing with drives 118/119

▶ Survival sense 122/123

SURVEYING THE MIND
▶ Brain maps 30/37

BUILDING THE BRAIN
▶ Food for thought 66/67

STATES OF MIND
▶ Anxious states 172/173

Dealing with drives

A tiny part of the brain is vital for controlling aspects as diverse as blood pressure, body heat, and thirst.

The hypothalamus, so called because it is under the thalamus, seems an unremarkable part of the brain: the size of a small, flat grape, it weighs only 0.3 percent of the brain's total. Yet it is the most important brain center dealing with homeostatis – keeping the body's internal conditions constant – and, from its position at the heart of the limbic system, is in overall charge of many autonomic and automatic body processes, such as body temperature, blood pressure, nutrition levels, and fluid balance. Nerve messages go to and from the hypothalamus and various control areas of the autonomic nervous system. It also exerts much of its control by links with the body's chief hormonal gland, the pituitary, just below.

But the hypothalamus is no "dumb autopilot." It is also the mediator between the unconscious autonomic functions in the brain stem and conscious awareness – the "higher" mental activities in the forebrain. You may be out walking, for instance, when suddenly you feel the need for a drink. This "need" is the hypothalamus, on behalf of its autonomic subsystems, breaking through into your conscious awareness and modifying your behavior to meet your body's basic needs.

Inputs arrive in the hypothalamus as nerve signals. They may be from the senses, especially sight, taste, and touch; from the limbic system or reticular formation in the brain itself (about basic needs and drives); and from internal organs such as the heart and intestines. It is also influenced by hormones circulating in the blood, and by concentrations of glucose, sodium, and other substances in the blood passing through it and in the cerebrospinal fluid just above it.

Inside the hypothalamus are several pairs of relatively discrete nuclei, or areas. Some tasks of the hypothalamus are assigned to a specific pair; others seem to be spread through several areas which communicate by nerve fibers so that they can work together as a single "center" for one main role.

The hypothalamus sends outputs as nerve signals to the motor (muscle-controlling) parts of the midbrain, the limbic system, and the various autonomic centers in the brain stem that control processes such as heartbeat, blood pressure, and urine production, as well as the glands producing saliva, sweat, and digestive juices. It also produces hormones – such as antidiuretic hormone (ADH), which makes the kidneys lose less water in urine, thus increasing blood volume and pressure – which pass hormones then to the pituitary.

Red nucleus

Basis pedunculi

Pons

Oculomotor nerve

A television studio is a hive of activity with many people working to produce a program. They are mainly autonomous, using their training and experience to get on with the job, but occasionally they must consult the studio manager, in overall charge of operations. The studio manager passes on high-level instructions as required and is the link with the actors and directors.

The hypothalamus is the body's studio manager. In overall charge of smooth running, it leaves routine and detail to the various homeostatic subsystems. It becomes active when something unusual happens. And it mediates between the service-type homeostatic activities and the brain's higher activities of awareness and conscious thought.

THE VITAL STALK

Hanging by a stalk from the hypothalamus is the pituitary gland, the chief controller and regulator of the body's hormonal processes. The nerve fibers and portal blood vessels in the pituitary stalk are the small but vital link between the body's two communication-control systems – nervous and hormonal. The hypothalamus makes its own forerunner "pro-hormones" in neurosecretory cells. These pass along the neurosecretory cell axons in the pituitary stalk for conversion into the mature hormone and storage in the posterior

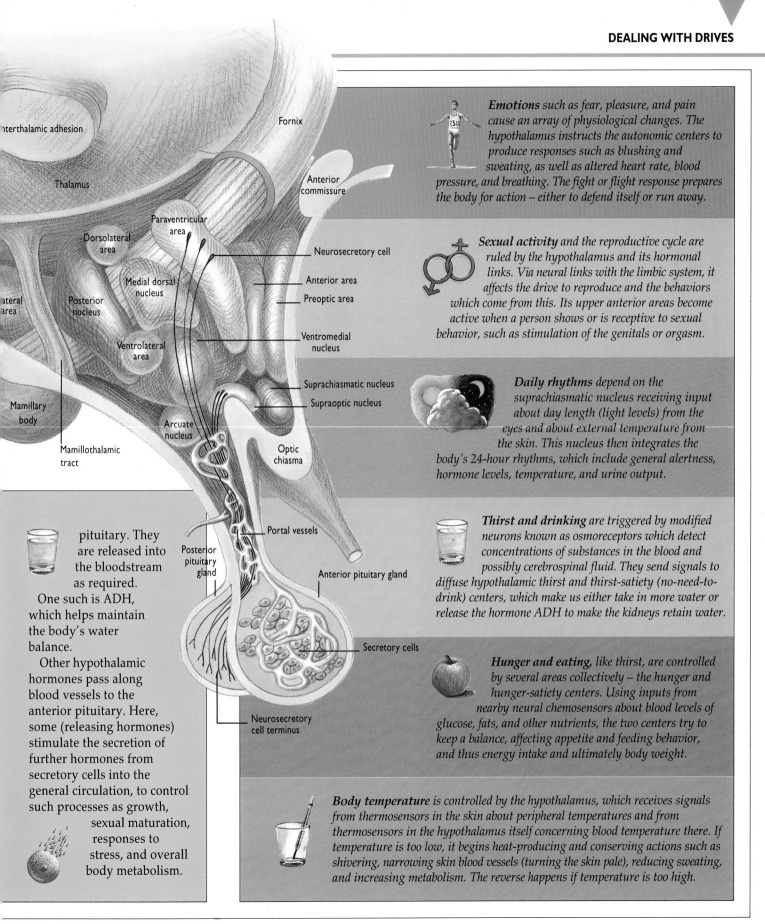

Interthalamic adhesion

Thalamus

Dorsolateral area

Medial dorsal nucleus

Posterior nucleus

Lateral area

Ventrolateral area

Mamillary body

Mamillothalamic tract

Paraventricular area

Arcuate nucleus

Fornix

Anterior commissure

Neurosecretory cell

Anterior area

Preoptic area

Ventromedial nucleus

Suprachiasmatic nucleus

Supraoptic nucleus

Optic chiasma

Portal vessels

Posterior pituitary gland

Anterior pituitary gland

Secretory cells

Neurosecretory cell terminus

Emotions such as fear, pleasure, and pain cause an array of physiological changes. The hypothalamus instructs the autonomic centers to produce responses such as blushing and sweating, as well as altered heart rate, blood pressure, and breathing. The fight or flight response prepares the body for action – either to defend itself or run away.

Sexual activity and the reproductive cycle are ruled by the hypothalamus and its hormonal links. Via neural links with the limbic system, it affects the drive to reproduce and the behaviors which come from this. Its upper anterior areas become active when a person shows or is receptive to sexual behavior, such as stimulation of the genitals or orgasm.

Daily rhythms depend on the suprachiasmatic nucleus receiving input about day length (light levels) from the eyes and about external temperature from the skin. This nucleus then integrates the body's 24-hour rhythms, which include general alertness, hormone levels, temperature, and urine output.

Thirst and drinking are triggered by modified neurons known as osmoreceptors which detect concentrations of substances in the blood and possibly cerebrospinal fluid. They send signals to diffuse hypothalamic thirst and thirst-satiety (no-need-to-drink) centers, which make us either take in more water or release the hormone ADH to make the kidneys retain water.

Hunger and eating, like thirst, are controlled by several areas collectively – the hunger and hunger-satiety centers. Using inputs from nearby neural chemosensors about blood levels of glucose, fats, and other nutrients, the two centers try to keep a balance, affecting appetite and feeding behavior, and thus energy intake and ultimately body weight.

Body temperature is controlled by the hypothalamus, which receives signals from thermosensors in the skin about peripheral temperatures and from thermosensors in the hypothalamus itself concerning blood temperature there. If temperature is too low, it begins heat-producing and conserving actions such as shivering, narrowing skin blood vessels (turning the skin pale), reducing sweating, and increasing metabolism. The reverse happens if temperature is too high.

pituitary. They are released into the bloodstream as required.

One such is ADH, which helps maintain the body's water balance.

Other hypothalamic hormones pass along blood vessels to the anterior pituitary. Here, some (releasing hormones) stimulate the secretion of further hormones from secretory cells into the general circulation, to control such processes as growth, sexual maturation, responses to stress, and overall body metabolism.

See also

INPUTS AND OUTPUTS
▶ Body links 80/81

▶ The long junction 82/83

▶ Keeping control 116/117

▶ Survival sense 122/123

SURVEYING THE MIND
▶ The working mind 16/17

▶ Inside the mind 20/21

▶ Brain maps 30/37

BUILDING THE BRAIN
▶ Building blocks 40/41

▶ Food for thought 66/67

FAR HORIZONS
▶ Male and female 148/149

STATES OF MIND
▶ Motivation 180/181

Rhythms of the mind

As natural products of evolution, the human body and brain are tuned to nature's rhythms and cycles.

The dominant cycle on Earth is the 24-hour circadian cycle of light and dark, the day–night cycle. Humans are mainly visual creatures, so it makes little sense to be up and about, or active and alert, during hours of darkness. The result is sleep, a state in which conscious self-awareness disappears, the brain takes time off for other tasks, and the body rests as its systems slow down and repair processes come to the fore.

Sleep is a major phase in the body's circadian biorhythms, which involve regular changes in hundreds of substances and processes, from levels of hormones and brain chemicals to body temperature, heart and breathing rates, and urine production. These mind and body rhythms seem to be internally generated, or endogenous, and persist even when people experience constant external conditions. They are largely based in two clumps of a few thousand neurons each, the suprachiasmatic nuclei (SCNs), in the brain's hypothalamic region just above the optic chiasma (where the optic nerves cross over). SCN neurons show their own circadian variations in electrical and chemical activity, and in the production of neurologically active substances – even when individual neurons are isolated in the laboratory. They are the nearest thing to a "body clock" thus far discovered.

The SCNs' built-in rhythms are synchronized, or entrained, with the external world in several ways. One is the light–dark cycle, detected by the eye's retina and communicated by nerve fibers from retinal ganglion cells to the SCNs. There are other entrainers whose mechanisms are much less clear; one such is the level of social interaction and arousal, which may communicate to the SCNs along fibers from the thalamus.

Nerve fibers from the SCNs communicate their information to other hypothalamic nuclei involved in many basic behaviors and body drives, and also along other more tortuous neural pathways, such as down the spinal cord and out to the cervical ganglia (nerve collections adjacent to the spinal cord in the neck) and then back up to the pineal gland at the base of the brain. The pineal makes several hormones and other chemicals, among them melatonin, which is produced mainly in darkness (light inhibits melatonin production) and which causes drowsiness.

EEG recordings of "brain waves" during sleep show regular variations in both frequency and amplitude. When asleep, the body passes from stage 1 (light sleep) down to stage 4 (deep sleep) and back up again, in periods of 80 to 120 minutes. These are punctuated by regular and increasingly long episodes of REM sleep (red bar), when the body becomes less active, except for rapid eye movement (REM), and dreams occur. The EEG traces from REM sleep are similar to those of wakefulness. During the day, alertness waxes and wanes in a similar cycle, thought to be related to the rise and fall in levels of neurotransmitters.

EEG changes　　　　　　**Levels of alertness over 24 hours**

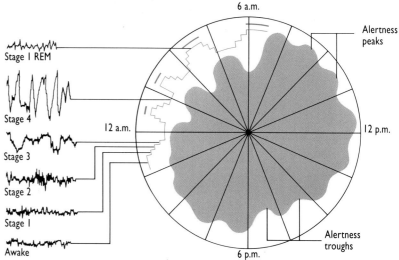

Stage 1 REM

Stage 4

Stage 3

Stage 2

Stage 1

Awake

6 a.m.

Alertness peaks

12 a.m.

12 p.m.

Alertness troughs

6 p.m.

Jet lag demonstrates that the human body was not designed to chase the Sun around the globe in an airplane at 625 mph (1,000 km/h). The aircraft's interior and comfort make the trip more or less bearable, but problems start on arrival in a different time zone. The brain and body say sleep, while the destination's night–day cycle, temperature variation, mealtime routines, and human activities upset internal biorhythms by saying stay awake. The results can be confusion, headache, memory lapses, drowsiness interspersed with sleeplessness, upset digestion, and reduced resistance to infection.

It takes most long-haul travelers between two and five days to reset their body clock. One recommended way of dealing with the problem is to try to conform to local time at once. But regular international trippers on short stays may try to stay on home time if at all possible.

FREE-RUNNING IN TOTAL ISOLATION

Frenchman Michel Siffre spent six months in the constant darkness, total silence and unvarying 72° F (22° C) temperature of Midnight Cave near Del Rio, Texas. He was wired to medical equipment on the surface which monitored various functions, including body temperature; brain wave, or EEG, traces; and heart activity. There was even a device that stored his daily beard shavings to study their hormone-influenced growth.

After about three weeks, Siffre's body had shifted to a 26-hour sleep–wake cycle. This became increasingly erratic, varying from 18 to 52 hours, although each seemed like a standard day length to the isolated cave-dweller. Over the six months, Siffre's average cycle was 28 hours. His mental processes and his manual dexterity in particular deteriorated steadily.

Because he was deprived of all environmental cues about time, from light–dark alternation to a clock, his body rhythm was allowed to "free-run" in its own self-generated cycle of sleep, wakefulness, and activity. Such studies help us to understand the body's internal clocks and rhythms, and they are of practical use to people such as astronauts and crew members on submarines.

See also

INPUTS AND OUTPUTS
▶ Feeling pain
110/111

▶ The primal sense
110/111

▶ Survival sense
122/123

BUILDING THE BRAIN
▶ Discovering transmitters
60/61

FAR HORIZONS
▶ Male and female
148/149

STATES OF MIND
▶ The resting mind?
160/161

▶ Feeling low
168/169

The sleep–wake cycle involves changing levels of neurotransmitters in the brain. There is twice as much acetylcholine present during both wakefulness and dreaming (REM sleep) as there is during non-REM, or ordinary, sleep. And while there is twice as much norepinephrine present during wakefulness as there is during non-REM sleep, there is only one-tenth the amount of it during REM sleep compared with non-REM sleep. The release of these chemicals is widespread throughout the cerebral cortex and therefore probably underpins overall changes in awareness. For instance, the variations in the alertness level over the course of a day (**left**) are thought to coincide with variations in the level of amine neurotransmitters (one of which is norepinephrine). Alertness rises with greater amine release and declines when levels are lower.

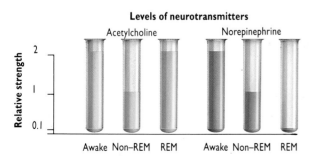

Levels of neurotransmitters

Acetylcholine   Norepinephrine

Relative strength: 2, 1, 0.1

Awake Non–REM REM   Awake Non–REM REM

Survival sense

With our superb senses and responses, we seem well equipped to cope with the modern world. But are we?

Just suppose that while walking along the road one day you came upon a lion. Surely you would be doomed. But wait; the human body in the modern form of *Homo sapiens* probably evolved on the plains of Africa, at least 100,000 years ago. Physically and physiologically, we have changed little in that time, and our distant ancestors must have encountered the occasional lion in their original home. In fact, if you pause to think about it, the human body is well suited to the situation. It is large and strong compared to most animals; its general senses are relatively sharp, especially sight. The eyes pick up the right range of electromagnetic waves, or light, to gain maximum information from the environment and discern every detail of the big cat's movements and behavior.

The body's other senses, such as hearing and smell, are similarly tuned in to important stimuli. Our intelligent brain can turn almost anything handy into a usable weapon. And the body's nervous and hormonal systems can quickly put its interior on "red alert." Nerve signals flash through the sympathetic nervous system, and epinephrine and associated hormones pour into the bloodstream. Almost at once, heart and breathing rates rise, blood pressure goes up, pupils dilate, and blood is diverted from the skin, digestive, and excretory systems to the muscles. All this happens fast and automatically, thanks largely to the brain.

At the same time, your conscious mind focuses on the lion. Your brain filters out irrelevant sights and sounds. Your conscious perception is super-aware, and your thought processes speed up. Your whole body

In a flight simulator a trainee pilot practices landing. All of his senses are working at their optimum as he concentrates on the task at hand. He watches the runway and the instrument displays, feels the plane's motions, and hears the engines' whine and warning buzzers. All of these mechanical and electronic systems are designed to stimulate and work with the body's senses. The brain copes with amounts of data coming along nerves from various sensory sources, such as sight, hearing, touch, and motion detection, that would swamp a supercomputer.

Modern humans cannot usually deal with stress by smashing its cause to pieces. The nagging colleague is often at a distance – on the other end of the phone. But the fact that we might dearly love to take physical action could reflect our ancient instincts and the body's fighting expectations.

TOO MUCH OF A BAD THING

Unrelieved stress can eat away at both mind and body, and the list of stress-related conditions grows ever longer. A mild heightening of autonomic nervous activity stimulates many body processes, such as acid secretion by the stomach in expectation of food. But when this state is maintained over days, months, and years, the excess acid erodes the stomach lining, and the result can be a gastric ulcer. Other stress-related conditions include headaches, breathlessness, asthma, heart flutters and angina, muscle twitches and aches, and many mental and emotional states such as anxiety, mood swings, depression, and schizophrenia.

Comparison of stress (life events)	U.S. Rating	Europe Rating	Japan Rating
Death of spouse	1	1	1
Divorce	2	3	3
Marital separation	3	5	
Jail term	4	2	2
Death of close family member	5		4
Personal injury or illness	6		5
Marriage		4	6
Pregnancy		6	

Some of the most stressful events are perceived differently in different cultures. The top six in the United States, for example, are not the same as in Europe or in Japan.

See also

INPUTS AND OUTPUTS
▶ The big picture 114/115

▶ Dealing with drives 118/119

SURVEYING THE MIND
▶ What is a brain? 12/13

▶ Inside the mind 20/21

FAR HORIZONS
▶ The evolving mind 126/127

STATES OF MIND
▶ Motivation 180/181

and mind are prepared for action, in a state of heightened readiness either to battle with the foe or to flee the danger. Then you see that it is just a stuffed lion, on its way to a museum. The result is neither fight or flight. Your nervous and hormonal systems gradually subdue the body's inner processes, trying to return to normal. This may be an extreme example, but the body's state of readiness is thwarted unnaturally like this time and again in the modern world. An argument at work, a traffic jam, an unreasonable neighbor, a family problem, unfair pressures from employers can all evoke the same sort of "primitive" reaction. The brain and body expect action as the response to stressful situations. But they rarely get it, and this can lead to both physical and psychological problems.

Stone Age humans coped with stress by taking action against its cause, be it a marauding leopard or rival humans from a nearby group encroaching on their territory. The human brain, nervous system, and senses were geared for survival in this type of situation. And the stone ax was one of the earliest tools used to aid survival.
Another more surprising, but extremely important, aid for general survival is pain. If we did not feel pain when we stuck our hand in a fire, for instance, we would leave it there and suffer burns which could reduce our chances of survival.

Far Horizons

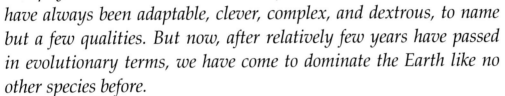

As a species, modern humans – Homo sapiens – have been around for a comparatively short period of time. We can look back over only about 5,000 years of recorded history. And evidence from various sources suggests that we existed for perhaps a couple of hundred thousand years before that. During that time we have changed little, both in our physical and in our mental capabilities. We have always been adaptable, clever, complex, and dextrous, to name but a few qualities. But now, after relatively few years have passed in evolutionary terms, we have come to dominate the Earth like no other species before.

It is the power of our mind that has enabled this. This power has given us the ability to control our environment. It has given us the means to learn and remember. And it has given us language. This, perhaps more than any ability, is the key to our success, for it allows us to communicate, passing knowledge down the generations so that it accumulates and lets us build on what has gone before.

*Left (**clockwise from top**): the developing mind in the growing child; getting a grip on interpersonal intelligence; drawing conclusions about autism; a superstitous response; memory has its high points. **This page (top)**: rising to the intellectual challenge of chess; (**left**) expressing ideas in a form of writing.*

The evolving mind

Due to our inbuilt intelligence, our numbers have increased and our achievements in all fields have reached extraordinary heights.

Back in the mists of time, about 20 million years ago, the first of the terrestrial apes evolved in Africa and inhabited the Miocene landscape. Perhaps 6 to 8 million years ago, this group split into two – the modern apes and the hominids. The first hominids are called australopithecines, or southern apes; the earliest fossil remains date back about 4.5 million years. Despite the name, they had some distinctly unapelike characteristics. For a start, they habitually walked upright, and their brains were larger than would be expected in an ape with the same body size.

Scientists believe that these creatures were the ancestors of modern humans, but their brain was only about a third the volume of ours. Over millions of years, brain size in succeeding species of hominids increased dramatically until the current figure was reached. So how did we get the big, efficient brains that give us our intelligence and ability to manipulate our environment?

Making and using tools was almost certainly a factor – bigger brains made for better tool makers, who were more successful still. Another suggestion is that standing upright reduced blood pressure and thus rate of flow of blood to the brain, which meant it was unable to lose enough heat. So the number of blood vessels there had to increase, which in turn led to more brain cells. Another theory links it with increasingly important social skills, especially the development of language.

The major design problem caused by increasing brain size is that the brain is an energy guzzler. Although it makes up only 2 percent of body weight, it uses 20 percent of the body's fuel in the form of glucose and oxygen. So a bigger brain demands more nutritious food – animal protein – and for our ancestors obtaining this required greater intelligence. And so came the ability to make better hunting weapons, to cooperate in hunting groups, and to communicate.

The Lascaux cave paintings *in the Dordogne in France were created by humans 25,000 years ago. The vivid use of representational and symbolic images shows that in their mental capacity these people were close to us.*

Australopithecines roamed Africa *from about 4.5 million years ago. One form was* Australopithecus robustus, *with a brain volume of about 30 in³ (485 cc). By 2.5 million years ago,* Homo habilis *had emerged. They made the first stone tools and had a brain of 40 in³ (650 cc). A million years later came* Homo erectus, *who made better tools, used fire, hunted big game, and whose brain was around 60 in³ (990 cc).* Homo sapiens neanderthalensis, *a variant of the human species, with an average brain volume of 100 in³ (1,640 cc), died out 35,000 years ago. Today's* Homo sapiens sapiens, *with a brain of 98 in³ (1,610 cc), originated 250,000 years ago and gradually replaced other forms.*

Australopithecus robustus　　　Homo habilis　　　Homo erectus　　　Homo sapiens neanderthalensis　　　Homo sapiens sapiens

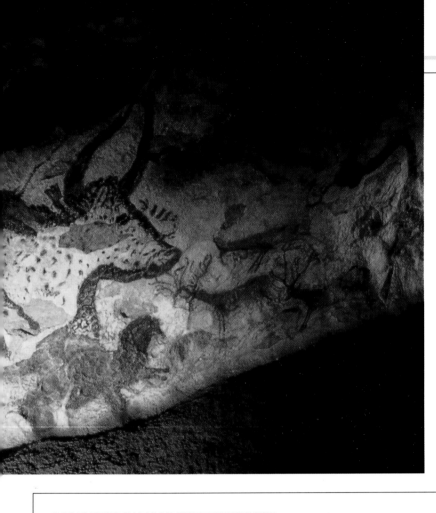

Our species survived by hunting and gathering, and the population was a constant 10 million until about 10,000 years ago. Then the agricultural revolution saw new farming techniques that generated more food. Numbers began to rise, and progress was made in other fields. Stone tools gave way to metal, settled communities allowed civilizations to grow, and the demands of trade led to written language.

Microchip

World population reached a plateau of several hundred million from the height of the Roman Empire until the Black Death, the plague that reduced the population by about one-third worldwide. The population recovered, and advances in science and culture eventually led to understanding about public health measures and food production. These, with modern medicine and mass immunization against infectious diseases, have caused numbers to skyrocket in recent decades.

Mass immunization

Population (billions)

See also

FAR HORIZONS
▶ Meeting of minds 134/135

▶ Word power 136/137

▶ Having ideas 140/141

▶ First thoughts 142/143

▶ Learning to be human 144/145

▶ Thinking together 150/151

SURVEYING THE MIND
▶ What is a brain? 12/13

▶ The working mind 16/17

▶ Brain maps 30/37

INPUTS AND OUTPUTS
▶ Survival sense 122/123

STATES OF MIND
▶ Being aware 158/159

▶ Motivation 180/181

COMMUNICATING INTELLIGENCE

Some researchers believe gossip was a major factor in boosting brain size. It has been found that the bigger the cortex in the brain of a primate species, the bigger the size of a group. And being part of a sizable group is an advantage – there is more protection, and larger hunting groups get more food.

Apes hold a group together by grooming, among other things. Language may have enabled humans to "mentally groom" many people at once by chatting with them. So those of our ancestors with the best language skills – and thus with larger cortices – enjoyed the benefits of a bigger group and were more likely to survive to pass on their large brain to further generations.

Roman Empire

Discovery of New World

Clean water

Stephenson's *Rocket*

Galileo

Modern Medicine

Computer Age

Agricultural Revolution

First civilizations

Flint tool

Iron Age

Black Death

Industrial Revolution

Scientific Revolution

Stone Age

Bronze Age

Renaissance

Middle Ages

0,000   5,000   1,000 550   500   450   400   350   300   250   200   150   100   50   0

Years ago

Conditioning the mind

Learning plays a vital role in survival, but there is more to memory than recalling responses to obvious dangers.

The first time you put your hand on a hot stove will probably also be the last. The pain from the heat is such a strong stimulus that it triggers a reflex response – you snatch your hand away – and you now associate a hot stove with pain and have learned never to touch one again. In the everyday world, such learned associations, or conditioned responses, exist in countless forms: a car driver brakes at a red traffic light; if a type of food makes someone sick, he or she will avoid it in future. But associations and responses can be learned that are not useful, including phobias and superstitions. A car wreck can leave behind a lifelong fear of cars; wearing a certain pair of socks when an important game of football is won can make someone want to wear those socks whenever a game is played.

Humans are not the only creatures that learn responses to stimuli. All animals, even including flatworms, are able to learn conditioned responses, because without them they would die. The first person to study this type of learning systematically was the Russian scientist Ivan Pavlov, who used dogs as his subjects. The work of Pavlov and others convinced some psychologists that all human behavior could – if broken down into small enough sections – be explained in terms

Superstitions can be explained as conditioned responses to certain situations. For example, things occasionally fall on people when they walk under ladders, so the link has been created that walking under ladders equals bad luck. Sportspeople and actors are especially prey to these false connections, which is why they often have a lucky item of clothing, or feel they have to go through some elaborate ritual before performing to bring them luck. But most people do not get hit by objects when they walk under ladders, so they have not learned to associate ladders with danger directly. Thus it seems that there is much to human behavior that we cannot explain in terms of simple conditioned responses.

of conditioning. Much of the evidence for this idea came from work done with animals. American psychologist B.F. Skinner, for instance, showed that it was possible to teach rats and pigeons quite complicated behavior, such as running through mazes correctly or even playing a version of ping pong.

The core idea behind Skinner's thinking was that all living creatures make small actions which have an effect on their surroundings. If one of these small actions, called an operant, has some sort of pleasant effect, the creature is likely to learn to repeat the action. The behavior is reinforced by the pleasant effect, in Skinner's experiments usually some reward such as food. Eventually a series of small actions that lead to pleasant effects can be shaped by reinforcement into more complicated strings of behavior. Thus a pigeon will put together a sequence of operants and play ping pong.

In fact, conditioning techniques, in the form of rewards and punishments, are used extensively in the human world, for example, to teach children how to behave properly. Skinner made the distinction between two types of reinforcers. For humans, primary reinforcers are things that are actually wanted, such as food, drink, sex, or rest; secondary reinforcers (the most obvious of which is money) are things that enable people to obtain the primaries.

See also

FAR HORIZONS
▶ The infinite store?
130/131

▶ Active memory
132/133

▶ First thoughts
142/143

▶ Learning to be human
144/145

SURVEYING THE MIND
▶ The working mind
16/17

BUILDING THE BRAIN
▶ Remaking the mind
62/63

INPUTS AND OUTPUTS
▶ Dealing with drives
118/119

STATES OF MIND
▶ The failing mind
176/177

HUNGER FOR LEARNING

Ivan Pavlov (1849–1936) was a Russian physiologist who, in 1904, won the Nobel prize for his work on digestion. While researching the amount of saliva produced by dogs, he noticed something intriguing. To begin with, his dogs would salivate only when they saw food, but they soon began salivating when they heard the sound of rattling dinner bowls, without any food being present. They had learned to associate the rattling of the bowls with food.

Over many years, Pavlov explored a large number of variations on this simple type of learning. It is known as classical conditioning, and it occurs as a result of association of a stimulus (in this case, rattling bowls) with an involuntary response (salivation) to a different stimulus (food). Pavlov linked food with lights and bells, and tested how close together in time the two stimuli had to be for the dogs to make an association. He also investigated what happened when he set up connections between unpleasant stimuli – electric shocks – and neutral objects such as lights and bells. He also discovered how to extinguish connections. Applied to humans, classical conditioning could explain why there might be a rush on the concession stand in a theater when a film depicting desert scenes was shown even though the temperature in the theater remained cool.

The lure, to some, of a slot machine can be explained in terms of a particular pattern of rewards for behavior. Psychologist B.F. Skinner found that rats and other creatures learned and then behaved in different ways depending on how he rewarded them for performing certain acts. In one experiment Skinner varied the number of times an action, such as pressing a lever, had to be performed before a rat was rewarded. He found that rats worked hardest and longest when they were rewarded at a rate of once for every five to seven actions. In another experiment, the time between the rats' actions and their receiving the reward varied unpredictably, and this often led to superstitious behavior. Humans find themselves in a similar situation when trying to dial a busy phone number; if, for example, they finally get through on several occasions after counting to 50, then they might repeat the pattern whenever they get a busy signal.

Humans are operating on a similar type of reward schedule when they gamble on a machine such as a slot machine. From time to time, the machine pays out a small amount of money, encouraging the gambler to go on playing. Whether or not this explains all the reasons why some people gamble is a matter of debate. For instance, how can we account for the fact that people still gamble when they know that in the long run they will lose? Clearly there is more to human life than the simple stimulus and response or conditioned response behaviors that a behaviorist such as Skinner would use to explain actions.

The infinite store?

Some memories are laid down in fairly precise regions of the brain, while others are spread right across it.

What were you doing yesterday evening? Such a question probably prompts a visual image, perhaps of yourself watching television, at the gym, or at a shopping mall. So is memory like a film library where all your experiences are stored waiting to be called up? The American researcher Karl Lashley believed it was and during the 1950s tried to locate the site of this library in rats' brains. He trained the animals to solve a maze and then removed different parts of the cortex to see if the loss of any area made them forget the solution. It did not.

Memory is more complicated than that. Many memories – how to ride a bicycle or the French word for cake – are not stored as images, so we have other sorts of memory. These are controlled by different parts of the brain. Studies on brain-damaged patients have shown just how specific these can be. One woman with damage to her left temporal lobe could remember the names for tools, but not for animals, for instance.

Our memory capacity is huge. Some people have a vocabulary of about 100,000 words, and ancient Celtic storytellers frequently knew 350 long epic poems by heart. Such feats, as well as ordinary everyday memory, rely on certain principles. The most important of these is association: the more links there are to other similar facts, the more easily a memory is recalled. So a number can be better remembered if it has a visual association, such as "legs eleven."

A model of the Eiffel tower will trigger many different memories in different people, even if they have never seen it, because it has reached the status of an icon. It will make most people think of Paris, but some may associate that with romance and moonlit walks by the Seine, while others might visualize other great structures and still others might remember a scene from a favorite film set in that city.

Memories are physically formed by strengthening the links between brain cells – the more links, the stronger the memory. There are around 100 trillion of these connections, or synapses, which explains our virtually limitless capacity, more than the largest supercomputer. Complex chemical changes at these links make certain connections more likely, and these are responsible for laying down memories.

A phone call may seem a simple act, but it can involve all the types of memory. The number may be held in the phonological loop for dialing, which requires non-declarative memory for motor skills. And without semantic and episodic memory, you would not be able to discuss the previous night's events.

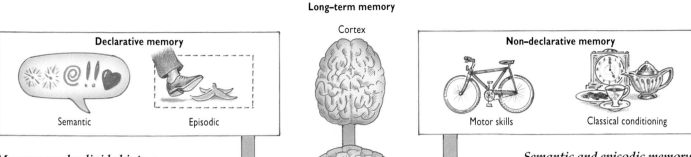

Long-term memory

Declarative memory

Semantic   Episodic

Cortex

Non-declarative memory

Motor skills   Classical conditioning

Hippocampus   Thalamus   Cerebellum

Working memory

A B C 1 2 3
Phonological loop

45 + 7 =
Central executive

Visual–spatial scratch pad

See also

FAR HORIZONS
▷ Conditioning the mind 128/129

▷ Active memory 132/133

▷ Word power 136/137

▷ Learning to be human 144/145

SURVEYING THE MIND
▷ Inside the mind 20/21

▷ Discovering through damage 24/25

▷ Brain maps 30/37

BUILDING THE BRAIN
▷ Remaking the mind 62/63

INPUTS AND OUTPUTS
▷ Guided motion 86/87

STATES OF MIND
▷ The failing mind 176/177

▷ Individual minds 184/185

Memory can be divided into a number of different types, *thought to reside in various regions of the brain. Both declarative and non-declarative memory are part of long-term memory. Declarative memory involves the use of the hippocampus for forming memories and the cortex for storage. It covers your memory of facts such as events and names which do not need to be repeated for them to sink in. For example, if you fell over on the way to work, you would remember it.*

Non-declarative memory covers motor skills, such as playing football or riding a bicycle. These memories are thought to reside in the cerebellum and do not seem to involve the hippocampus. They require rehearsal and patience for you to learn the skill.

Semantic and episodic memory *are subsections of declarative memory. Semantic memory is the knowledge of facts – language and concepts – which the brain files in categories and which seems to involve the left temporal lobe. Episodic memory is of an event in one's life and everything about it, including emotional reactions to it.*

Classical conditioning, *along with motor skills, is part of non-declarative memory. Our desire for food at a particular time of day – regardless of whether we are hungry or not – is one example of such conditioning.*

Short-term, or working, memory *is also divided into categories – of which there are three main ones. They all work together to keep material in the mind for a short time, and allow us to manage such tasks as putting together a sentence. The first of these is the phonological loop, which enables us to remember sequences of approximately seven digits, letters, or words.*

The second is the visual–spatial scratch pad. It is like a sort of inner eye, which receives and codes data into visual or spatial images. It comes into play when we need to remember where we were on a page when we start reading a book again. The central executive is the last of these elements. Its function is to help with such tasks as reasoning or doing mental arithmetic, and as such it is rather like the RAM of a computer.

CUT OFF FROM THE PRESENT

✚ A man had suffered from severe epileptic fits since he was 16 years old, but all drug treatments had failed. In an attempt to control the condition, his doctors decided to operate on the area of his brain that was affected, the area below the temporal lobes. This was in the 1950s, when brain surgery for all sorts of brain disorders was far more common than it is now.

In one sense, the operation was a success – the man was cured of his convulsions. In another, its effects were devastating: as a result of the surgery, he was unable to form any new memories. Any conversation he had was forgotten after a few minutes, as was the person

he spoke to. He could not remember where he lived and could read a book several times without any sense that he had read it before. Yet he did not lose skills such as language, and his memory of events some two years before the operation was normal. What is more, he taught himself to draw, showing that he could still memorize motor skills, but he had no memory of learning to do so.

The importance of this case is that it showed for the first time that the hippocampus – which was damaged on both sides of his brain by the operation – is involved in laying down memories, but it does not appear to be the place where they are stored.

Active memory

Just as vision is an active creative process, so memories seem to be remade every time we call them to mind.

Generally, our memory serves us well. We remember people's faces and names, we know what we did yesterday or last week; but it does not work like a computer disk. It seems that rather than storing memories, we re-create them each time we recall them, and that means that they can change. This is why eyewitnesses are notoriously unreliable. We do not record the events like a camera; instead, what we remember can be affected both by what we expect to see and by the questions we are asked about it. Many people expect bank robbers to carry guns, so if shown a film of an unarmed robbery they will often say that guns were used. Likewise, asking witnesses about a "car wreck" as opposed to an "accident" makes it more likely they will come up with false memories of broken glass.

So called false memories have also featured in cases of alleged sexual abuse in childhood. Numerous people in therapy claim to have uncovered long-buried memories of abuse. Some claim that certain of those memories, which may seem totally real, have been influenced by the therapist's suggestions.

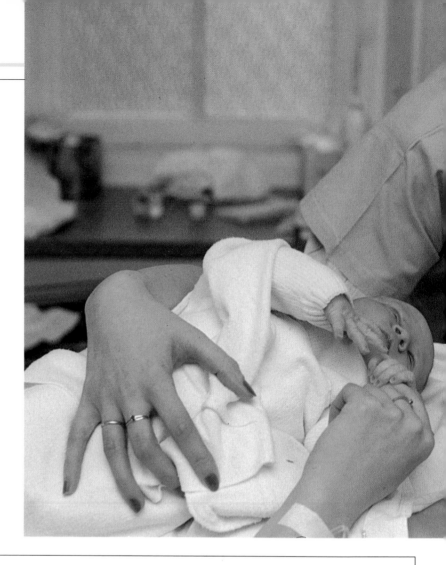

LOOKING FOR THE SEAT OF MEMORY

One of the apparently strongest pieces of evidence that memories are stored in snapshot form came from the work in the 1940s and '50s of the Canadian neurosurgeon Wilder Penfield, who developed techniques for treating epilepsy. Since epileptic fits are caused by a diseased patch on the brain, he worked to find that patch so it could be removed. To do this, he removed the top of the patient's skull, using only a local anesthetic (to numb the scalp, the brain itself feels no pain), and stimulated the exposed brain with electrodes. The idea was that when he found the right patch, the patient would experience the feelings that normally preceded an attack.

Not only did Penfield locate the site of the disease in his patients, he also believed that he had found the seat of memory – in the temporal lobe. When this area was stimulated, 40 out of about 1,000 patients reported vivid flashbacks – a fragment of a tune, a child calling, being in a room. All were marked by a dreamlike quality with no sense of time or location. He believed that this was evidence that our memories are stored in one place in a complete and recoverable form. Later researchers pointed out, however, that few of the patients recalled actual memories; so perhaps the position in the brain of the seat of memory is a mystery still to be solved.

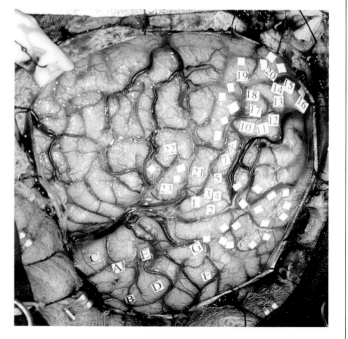

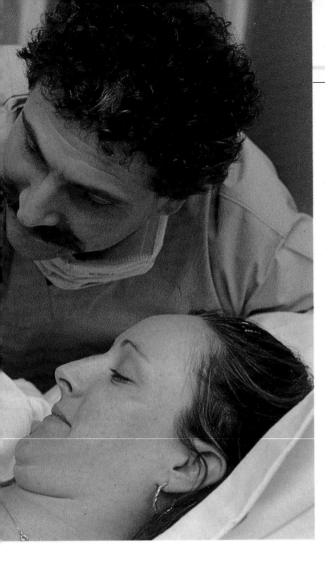

The birth of a baby is an event that no mother – or father, if he is present – will ever forget. Every moment seems etched on the mind (although women seem to be selective enough about memories of the pain they experience to be prepared to do it again). But all of us, whether we are parents or not, have similar personal moments that stand out especially vividly, and which we seem to remember accurately rather than distorting with each recall. They are known as flashbulb memories and are apparently "burned" into our brain because the original experience was so powerful.

Something similar seems to happen to the victims of post-traumatic stress disorder, who often have vivid flashbacks to a terrible event – usually one that threatens their own life or that of a family member or close friend – for years after the original incident. Unlike normal memories, these remain exceptionally strong unless treated with therapy soon after the event. The symptoms of this disorder include recurring nightmares in which sufferers relive the event, feelings of guilt at having survived, sleeplessness, and a profound sense of detachment.

YOU MUST REMEMBER THIS

Imagine being able to remember huge strings of random numbers with the ease that you remember your mother's name. In 1920s Moscow, a newspaper reporter could do just that, not because he had a particularly good general memory, but because he was able to think up mnemonic tricks incredibly quickly. His technique was to visualize a familiar street or place and then associate the list of numbers or even objects that he wished to recall with particular aspects of that scene. To remember the numbers, he simply had to imagine himself in the scene he had created (often many years before) and find the places he had hidden them.

A slightly different technique has been used by a British mnemonist, who associates numbers with letters, making words, and from the words creates a – sometimes bizarre – picture. For instance, faced with 36 random numbers to remember, he breaks them down into groups of 6. Using a special conversion table that he has memorized, he first gives the group a number, so that he knows

where it comes in the sequence. If he were trying to remember the fourth group, this would give the letter R, which he makes into a word by adding vowels – "ray," which he remembers as a ray of light. If that group were, say, 351639, he would get the letters MLTChMP. By breaking this down, he gets MLT and ChMP, which with the addition of vowels might give the words "mallet" and "chimp." The whole gives the unlikely image of a chimp smashing a light with a mallet, one he is unlikely to forget.

In everyday life we rarely have to perform such feats of memory. But we can use similar tricks – ideally using unusual visual images, since memory works best with them – to recall more useful things. For instance, to make the name Barbie Carson memorable, you might conjure up an image of a Barbie doll. Then for Carson, try a picture of a car driving toward the setting sun. You will not forget a doll at the wheel of a bright pink Cadillac at sunset.

A knotted handkerchief is often used to jog memory. When you find a knotted handkerchief in your pocket, it reminds you that you have to remember something – which will usually then come back to you. Failing that, leave yourself a note – if necessary written on your hand.

See also

FAR HORIZONS
▶ Conditioning the mind
128/129

▶ The infinite store?
130/131

▶ Having ideas
140/141

SURVEYING THE MIND
▶ Inside the mind
20/21

▶ Probing the mind
26/27

BUILDING THE BRAIN
▶ Remaking the mind
62/63

INPUTS AND OUTPUTS
▶ Active vision
94/95

STATES OF MIND
▶ Emotional states
154/155

▶ The unique self
156/157

▶ The failing mind
176/177

133

Meeting of minds

Across the animal kingdom, one of the prime abilities of the mind is communication with other minds.

Walk into a wood, field, or jungle and listen. What you hear is communication – a cacophony of chirps, calls, whistles, and grunts as animals advertise for mates and warn about the approach of predators. Birds stake out territory with song, monkeys grunt and cough to make arrangements with other members of the troop. And enter a place where humans congregate and you will hear a hubbub of talk, see a flurry of gestures, and observe myriad postures and expressions.

You have only to look at your own dog or cat to understand how much can be conveyed without speech. Your dog greets you with a welcoming bark and enthusiastic wagging of the tail when you return home; but a stranger who is perceived as a threat is met with a snarl and a curled lip. A cat carries its tail high over its back in greeting and purrs with pleasure when it is stroked. All of these are unmistakable signals, conveyed by body language and sound. But by far the most sophisticated form of communication is human language. While nearly all animals have only a limited and fixed number of calls and signs, language is almost infinite in its possible combinations.

Speech was impossible until about a million years ago because production of the complex sounds of language requires more space in the throat than our primate ancestors had. We acquired enough room by developing a longer neck and a lower larynx; we also produced a more rounded tongue – modifications that permitted the development of speech. But language is much more than making sounds, it requires the ability to use symbols and handle abstract ideas.

Primates are now considered to be more sophisticated communicators than was previously thought, and it appears that they can learn to use symbols and signs. So it seems that building on this ability, early humans at about the time of *Homo erectus* 1.6 million years ago developed sign language as a halfway house to speech. Not only had their brains become larger, as is evident from skulls of the time, but they were also bigger on the left – the side that controls language. In addition, they had developed an opposable thumb, which made holding tools and making hand signs easier. Language proper eventually emerged about 100,000 years ago or possibly even earlier.

A girl sitting with one leg propped on her knee and hands clasped behind her head is giving out strong non-verbal signals. Ask a group of people what message she is conveying and they will agree that she exudes arrogant self-confidence.

Indeed, her attitude is one that most people will recognize as that of a rather smug, over-confident individual, who will expound upon her own abilities and achievements to the boredom of all around her. Conversely, although it may be easy to read the body language of another person, we are often not conscious of it in ourselves. But some people learn to control their body language. Poker players, for instance, who do not want to give any indication of how good or bad their cards are, rigidly govern any giveaway non-verbal clues. Hence the expression "poker face."

CHIMP SIGNING

 In the 1960s it was announced that after five years of intensive training a chimpanzee called Washoe had learned 132 signs and could indicate general classes of objects such as "dog" or "bird." Recent research seems to indicate that young pygmy chimps, or bonobos – the type found in the rainforests of Zaire – can learn a similar number of signs fairly quickly. One bonobo, Kanzi, was able to understand 150 English words after only 17 months. Like a child, he learned language by experience, rather than by special training, and by the age of nine was able to comprehend and carry out quite detailed verbal instructions. Although incapable of speech, he was also able to express his feelings and wishes using a keyboard with 256 signs (**left**). But despite the abilities of some specially trained chimps, they lack a skill that is unique to humans. While chimps can manage to express their needs, for instance by signing that they want a banana, they cannot express ideas.

Researchers studying vervet monkeys in the wild have found that they have an extensive and complex range of calls and signals, more than was previously thought. There are several specific types of calls and signals that youngsters need to learn. One set warns of nearby predators and even distinguishes between leopards, snakes, and eagles, while the other involves a number of complex social situations. From such beginnings language may have developed.

See also

FAR HORIZONS
▶ The evolving mind
126/127

▶ Word power
136/137

▶ Parallel minds
138/139

▶ Milestones of the mind
146/147

SURVEYING THE MIND
▶ Comparing brains
18/19

INPUTS AND OUTPUTS
▶ Sensing sound
96/97

▶ The mind's ear
98/99

STATES OF MIND
▶ Talking cures
164/165

Non-verbal communication has always played an important part in transmitting messages. Researchers studying primates can tell simply from posture which of the animals is dominant or submissive, since junior monkeys appease their seniors by making a gesture which says, "Look, I'm not a threat." And the first serious research on human body language found that we, too, often use it to express rapport or to signal our state of mind. Films of conversations revealed that those who were listening often unconsciously moved their hands or head in time to the rhythm of the speaker's speech. Similarly, when people were getting along, they often sat or stood in the same way, and when they were angry or bored, it was evident from their posture, even if their words appeared to be placatory or polite.

With arms and legs crossed, *this girl conveys that she is on the defensive and probably nervous as well – an impression reinforced by her hidden hands. Perhaps she is shy, feels threatened, or is having to listen to an arrogant bore.*

Nose rubbing or touching, *usually with an index finger, which may be transmuted into rubbing an ear or an eye, generally signifies that the person is doubtful what his response should be to an awkward question or situation. It may also be accompanied by other negative gestures such as wriggling on the chair or turning sideways.*

Word power

The fact that you can read and understand the words on this page means that you must be a human being.

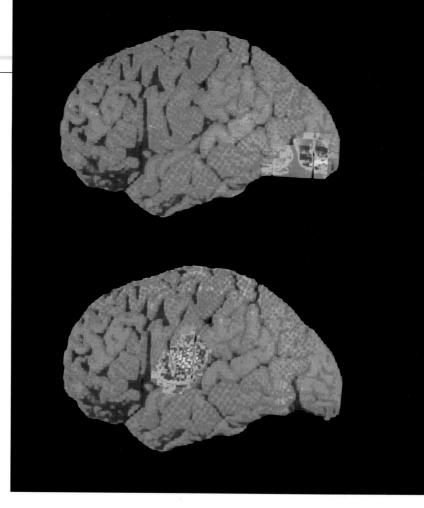

All human societies, however primitive their technology, have highly developed language, which they use to share ideas, plan complex activities, and teach their children. But because we learn to speak automatically, it is sometimes not appreciated what an amazing achievement language is. We not only have tens of thousands of words to describe things and ideas, but we also have grammar, which is one of the features that distinguishes human from animal communication.

Grammar is the complicated and elaborate set of rules for putting words together so that they convey meaning. Even the experts find it difficult to describe grammar fully, yet by the age of two, English-speaking children have, for instance, worked out the rule for making a word plural and can soon talk in reasonably grammatical sentences. For this reason, most researchers now believe that humans are born with an innate ability to learn language, just as we arrive with the

Written language first emerged in Mesopotamia about 5,200 years ago. Marks made on clay tablets with a reed pen recorded such things as grain accounts and land sales. Crude drawings of familiar objects such as human heads or animals were also used to convey ideas. Gradually, the pictograms became more stylized, until words were represented by symbols. The pictograms at the top are from about 4,400 years ago, while the symbols, called cuneiform, were made 2,650 years ago. How the signs sounded is shown at the bottom of each column.

*Language regions in the brain are shown by PET scans which detect active areas. The images of the left-hand side of the brain show the area activated by vision, including reading (**top left**); by hearing (**top right**); by speech (**bottom left**); and by the high activity generated by thinking about words and speaking them (**bottom right**).*

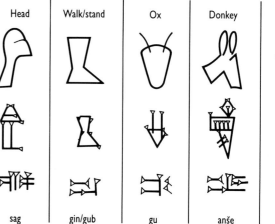

Head	Walk/stand	Ox	Donkey	Day	Water
sag	gin/gub	gu	anše	ud	a

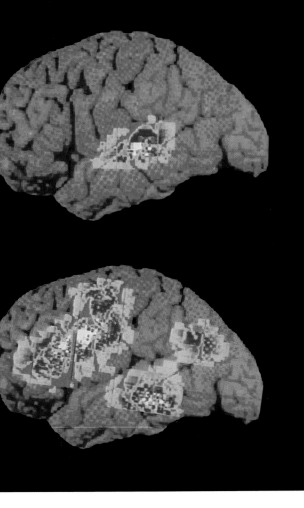

See also

FAR HORIZONS
▶ The evolving mind
126/127

▶ The infinite store?
130/131

▶ Meeting of minds
134/135

▶ Parallel minds
138/139

▶ First thoughts
142/143

▶ Milestones of the mind
146/147

SURVEYING THE MIND
▶ Discovering through damage
24/25

▶ Brain maps
30/37

INPUTS AND OUTPUTS
▶ The mind's ear
98/99

STATES OF MIND
▶ Talking cures
164/165

▶ The mind adrift
170/171

A TIME TO TALK

⚕ A boy of about 10 was found wandering in the wild. He appeared healthy and had been cared for by wild animals. He was taken in, but despite persistent efforts, he could not be taught to speak, even though there was nothing wrong with his vocal apparatus. It seems that unless children are exposed to speech before the age of about eight, they are unable to learn a true language.

Normally the process of learning language is marked by the same type of regular stages of development found in other pre-programmed abilities such as learning to walk. By the age of six months, on average, babies are constantly babbling, making all the sounds that are used in human languages, including some of the clicks that are a feature of a number of African languages. At a year, they can manage a few words, and by two years they are using two-word combinations such as "mummy ball" that show a grasp of the distinction between subject and object. Once children learn a word like "ball," they often apply it to other objects with the same function – such as bouncing – or the same form – round. What is more impressive is that they then make corrections without having to be specifically told. By the time they are four to five years old, most children have a vocabulary of several thousand words and have mastered the basic rules of grammar and syntax.

ability to see. Both can, however, be shaped by the world in which we find ourselves.

In places where adults – usually immigrant workers – who do not share a common language are living together, their children will spontaneously develop a pidgin language made up of bits from all their parents' languages. What is remarkable about these pidgin languages is that among second-generation speakers, clear and specific rules emerge completely spontaneously.

Language is generally processed by the left side of the brain, and damage in certain areas can lead to the loss of specific aspects of ability. For example, patients in whom part of the left frontal lobe, known as Broca's region, has been destroyed can understand language but cannot speak fluently, while damage to Wernicke's area, nearby in the temporal lobe, produces speech that is fluent but virtually meaningless.

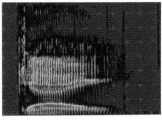
Sound: Gay

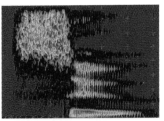
Sound: Sue

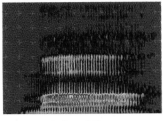
Sound: Ma

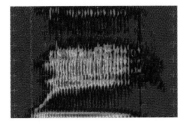
Sound: We

Voices make a combination of noises as shown in these images where sounds have been plotted according to their frequency (pitch) and volume (the lighter the color, the louder the sound). The brain *analyzes the complex sounds of speech, recognizing not only the content, in terms of grammar and the like, but also who is doing the talking by the distinct set of noises generated by each individual.*

Parallel minds

Reading these words probably makes the left side of your brain work a little harder than the right.

We each have a brain in two halves. Almost without exception, every part of the brain on the left is matched by a more or less mirror image part on the right. This applies to animals, too, and in many creatures – ourselves included – each side of the brain controls the opposite side of the body.

Humans are thought to be unique, however, in that opposite structures in the cerebral cortex, the thin layer of gray matter on the outside of the brain, have become specialized, with some functions not mirrored left to right. In most people, the left side, which is slightly larger, controls language and logical activities, and the right controls spatial activities and tasks with a more emotional element. The essential difference between the two halves seems to be not so much the type of information that has to be processed – words versus pictures – but how it is dealt with. For instance, only the left side is involved in reading a technical manual, but reading folk tales, with their strong mythic and symbolic content, requires both sides to work together.

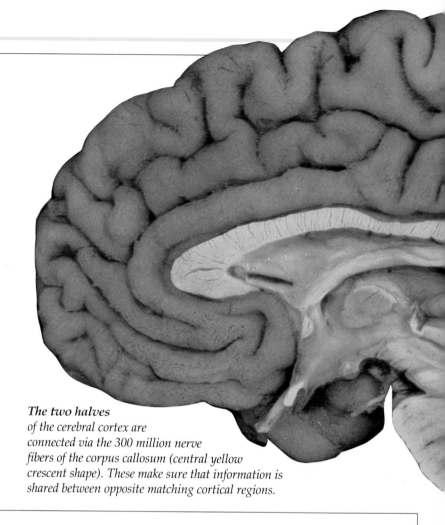

The two halves
of the cerebral cortex are connected via the 300 million nerve fibers of the corpus callosum (central yellow crescent shape). These make sure that information is shared between opposite matching cortical regions.

MUSIC IN MIND

Studies of musicians reveal that brain dominance can change as a result of learning and experience. For instance, non-musicians focus on the overall contours of the melody when listening to music and tend to use the right side of the cerebral cortex. But professionals have learned to be more analytical about it and so use the left side to break the music into its component parts.

The effect of brain damage on musical ability is also a good example of the way the two halves specialize. Damage on the right causes singing to be grossly disturbed and all sense of melody to go; but rhythm, which involves a repetitive sequence, remains. Left-side damage, however, which the French composer Ravel (1875–1937) suffered, causes problems with naming songs or recognizing written music. Similarly, recent studies with PET scans have shown that reading music involves left brain areas that are near, but separate from, those involved in reading.

See also

FAR HORIZONS
Meeting of minds
134/135

Word power
136/137

Having ideas
140/141

Male and female
148/149

SURVEYING THE MIND
The working mind
16/17

Discovering through damage
24/25

Brain maps
30/37

BUILDING THE BRAIN
Recovering from damage
64/65

INPUTS AND OUTPUTS
Levels of seeing
92/93

Sensing sound
96/97

The mind's ear
98/99

WHEN THE LEFT HAND DOESN'T KNOW WHAT THE RIGHT HAND IS DOING

Given a pencil to hold in his right hand, which is hidden from him, a man can describe it perfectly well. But ask him to do the same thing when the pencil is in his left hand, and he cannot describe it at all. Using his left hand, the man can make a rough copy of a square, but is unable to write the word "Sunday" – "SA" is all he can manage. With his right hand, by contrast, he can write the word perfectly well, but the square is represented as five parallel lines above one another in a column.

The man suffers from severe epilepsy and has had the connection between the two halves of his cerebral cortex, the corpus callosum, cut as an extreme treatment to control his seizures. Each side is thus forced to function on its own. So the right cortex, which controls the left hand and receives sensory information from it, can neither give the pencil a name nor write a word, as it is specialized in spatial tasks. But it can instruct the left hand to draw a square. The left side, which deals with the right hand, can give the pencil a name and can write the word "Sunday" as it is specialized in language. But it cannot tell the right hand to draw a decent square.

*The left side of the brain controls the right of the body and receives sensory information from it. Similarly, the right controls and receives data from the left. Thus the left motor cortex (**below**) controls*

The ability to use and understand language and to perform fine motor skills is usually located in the left side of the cerebral cortex, as are the musical abilities of timing, sequencing, and rhythm. So, too, are logic, mathematical abilities, and the analysis and shaping of ideas.

The right side is, in most people, better at dealing with complex visual patterns, perspective, and spatial matters. It is also thought to be more involved in emotions and in insightful thinking. The musical elements of tonal memory, quality of sound, melody recognition, and intensity are also handled here. The right side is responsible for

Motor cortex
Somatosensory cortex
Angular gyrus
Broca's area
Left primary visual cortex
Wernicke's area
Left auditory cortex

recognizing faces and interpreting their expressions. Although each half of the cortex has its functional specializations, there is a constant flow of data between them via the corpus callosum. There is some evidence that the callosum is thicker in women, leading to speculation that the female brain may be better integrated.

the motions of the muscles in the right side of the body, and the left somatosensory cortex receives touch signals from the right of the body. In addition, the left primary visual cortex deals with what is in the right of the visual field, and the left auditory cortex deals with sounds from the right ear.

Three areas on the left of the cerebral cortex are known to have specific language functions: Broca's area (which deals with grammar and articulation); Wernicke's area (which covers sense and understanding); and the angular gyrus (which turns images into words).

For the overwhelming majority of right-handed people, the left cerebral cortex does indeed handle language and logic, and the right spatial tasks. But in left-handed or ambidextrous people, the functional specialization can be different. Sometimes there is a straight swap between the hemispheres and their specialized functions, so a left-hander's right brain might deal with language and logic. But sometimes functions are shared equally between left and right.

Having ideas

From daydreaming to solving problems to having an original thought, the brain can do it all.

The phenomenal complexity and flexibility of our brain allows us to perform extraordinary feats. Our 100 billion neurons give us the processing power of a vast computer. This allows us to deal both with mundane everyday problems and with mentally demanding tasks such as understanding and using complicated mathematical concepts. The brain's higher functions give rise to language, technology, and the arts; yet, although we all know what it means to be creative or intelligent, understanding how we do it has proved to be difficult.

Stephen Hawking theorizing about black holes, Picasso transforming a bicycle seat and handlebars into a bull's head, and a chess grand master are all operating at a high level of intelligence. But they are also doing things that require very different mental skills. And measuring mental skills is one of the most controversial areas of psychology. One way is the IQ (intelligence quotient) test, which uses multiple-choice questions to test how good people are at verbal skills – reading, writing, vocabulary – and spatial ones – arranging blocks, detecting patterns. It appears to reflect a basic division, into two types, in the way we think. The first is known as convergent and refers to organizing, analytical skills. The other is divergent and includes the brain's ability to make unexpected connections. Artists such as Picasso tend to be divergent while scientists and chess players are more convergent.

Much of our high-level thinking seems to be done in the frontal lobes of the brain, but recent PET (positron emission tomography) scans, which show how much energy the brain uses, have come up with some surprising results. Not only do bright people seem to use less energy when solving problems than those of average intelligence, but male and female brains seem to tackle problems in a different way and using different areas. Such scans may eventually resolve the debate over whether there is one central quality to intelligence or several independent ones.

*It takes a leap of imagination to create a work of art from unlikely materials. The bull (**left**) was made from recycled junk found on cattle ranches in New Mexico by artist Holly Hughes.*

Just how or why some people are able to come up with genuinely original ideas is still not fully understood, but it seems to be linked to personality. Some researchers associate it with psychoticism, while others connect it with manic-depression, or bipolar disorder.

Creativity does not feature in IQ tests because it is not easy to measure – a person's ability to create something new is hard to assess objectively in the same way as other forms of intelligence.

As far as the location of creativity in the brain is concerned, the right side seems to be most concerned with producing new images or connections, while the left is essential for evaluating and shaping them.

See also

FAR HORIZONS
▶ The evolving mind
126/127

▶ Parallel minds
138/139

▶ Milestones of the mind
146/147

SURVEYING THE MIND
▶ The working mind
16/17

▶ Inside the mind
20/21

▶ Brain maps
30/37

INPUTS AND OUTPUTS
▶ Active vision
94/95

STATES OF MIND
▶ The unique self
156/157

▶ Abnormal states
162/163

▶ Feeling low
168/169

▶ The mind adrift
170/171

IDEAS ABOUT INTELLIGENCE

The IQ (intelligence quotient) test is the best-known attempt to rate intelligence on a single scale, although it does not deal with some forms of intelligence. Psychologist Howard Gardner broadened out the idea of IQ with his multiple component theory, in which there are a number of types of intelligence (**right**). Some argue, however, that including specialized abilities like music and mathematics means his model does not apply to everyone.

Linguistic intelligence is the use of language in fundamental aspects such as writing, reading, and understanding speech. This is best exemplified by the creative use of language required, for instance, in writing poetry.

Personal intelligence has two forms: interpersonal intelligence – used in interactions with others; and intrapersonal intelligence – concerned with understanding and knowing yourself.

Logical-mathematical intelligence is of the type used in numerical calculation, arithmetic, and logic. It involves the ability to manipulate quantities and is separate from the abilities needed for language and music. Work such as Einstein's requires this type of intelligence.

Bodily kinesthetic intelligence is the ability to use the body expressively and skillfully, especially in fine motor control of the hands. It is used in sports, dancing, and simple everyday movements involving dexterity. Someone with a high score in this type of intelligence might perform delicate surgery well.

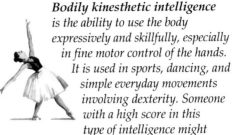

Musical intelligence is used in appreciating, performing, and composing music. Composing has a logic of its own, quite distinct from that of language. It can be one of the most striking early talents – the composer Mozart was a child prodigy.

Spatial intelligence involves being able to perceive the shape and relative position of objects. It is evident in the ability to design and build things, from tables to planes. It also involves being able to find your way around and is important in art.

Solving chess problems involves the use of spatial intelligence, logical mathematical intelligence, and phenomenal feats of memory.

First thoughts

Babies' brains are very flexible, and they develop rapidly after birth in response to the world around them.

A new baby is virtually helpless, but it has a number of reflexes to help it feed and bond with its mother. For instance, the cries it makes when distressed – which usually bring a rapid response – are just one method it can use to make sure that it survives. And recent research has shown that even before birth the fetus has a number of capabilities. It does regular physical exercise in the womb, at first jerkily, but later with more coordination; it also shows disgust if the amniotic fluid that surrounds it is made bitter. Later, the fetus will be startled in response to a sudden noise and will blink when a bright light is shone on its mother's stomach. If it regularly hears a piece of music while in the womb, it recognizes and responds to it for several weeks after birth.

Some things can interfere with normal, healthy development. If a mother contracts certain diseases or takes drugs, the baby can be brain damaged, and there is evidence that intense and unhappy emotional states in the mother can also have an effect.

BORN DAMAGED

The pediatrician looking after a newborn boy noticed that the infant was small for his age, with an unusually flattened face. As the boy grew up, he became hyperactive and also showed signs of being mentally retarded.

His symptoms were due to fetal alcohol syndrome, the third most common form of birth abnormality after Down's syndrome and neural tube defects such as spina bifida. The boy's mother was an alcoholic, and since mother and baby share a blood supply, drugs or deficiencies in the mother's blood can harm the baby.

The fetal brain is particularly vulnerable between 8 and 16 weeks after conception, when the brain is growing at a remarkable rate. One estimate is that 250,000 neurons are being generated every minute.

Other threats to a fetus are rubella (German measles) in the first two months of pregnancy, which can cause the baby to be retarded or deformed, and spina bifida, linked to low levels of folic acid in the diet. And there is some evidence of a link between flu during pregnancy and a slightly increased risk of schizophrenia.

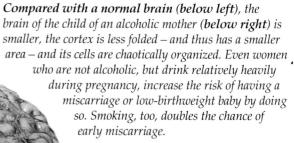

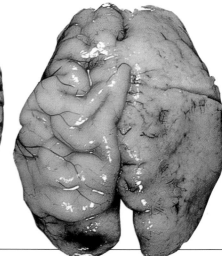

Compared with a normal brain (below left), the brain of the child of an alcoholic mother (below right) is smaller, the cortex is less folded – and thus has a smaller area – and its cells are chaotically organized. Even women who are not alcoholic, but drink relatively heavily during pregnancy, increase the risk of having a miscarriage or low-birthweight baby by doing so. Smoking, too, doubles the chance of early miscarriage.

Two inborn reflexes help a newborn to feed. The rooting reflex makes the baby turn its head toward a touch on one side of the mouth to find something to suck. It immediately begins to suck vigorously – the sucking reflex.

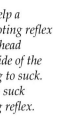

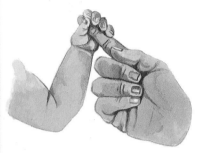

The grasp reflex is instigated by pressure in a baby's palm. For a few days after birth, the grasp is so strong that a baby can support its weight by holding its mother's finger. The reflex fades by four months, but may once have ensured survival.

The walking or stepping reflex can be triggered in a newborn baby by holding it upright and tilting it slightly forward with its feet touching a firm surface. The baby then makes steplike movements.

This reflex tends to disappear by the age of two months. It is thought to become impossible as the baby gains weight because its legs become too heavy for the muscles to make them move.

True walking, usually preceded by a set order of actions including crawling and supported standing, does not happen until about 12 months, although the age does vary considerably.

For reasons of survival, a newborn baby is a totally egocentric creature. The infant's world is small and revolves around satisfying its immediate needs for food, warmth, and care through its mother, since it is unable to look after itself. Thus the baby can focus on objects 10 inches (25 cm) away, about the distance to its mother's eyes when it is feeding. And it seems programmed to respond to faces. Given a number of colored disks to look at, a baby will spend the longest time staring at the one with a crude representation of a human face on it. Within a few hours of birth, a baby can follow a moving light and can recognize sounds that it heard while in the womb.

It is not known for sure what a baby experiences of the world around it. Some say the world must seem a confusing place after the peaceful darkness of the womb. Others believe that perception is probably quite coherent, but still simple and less selective. And as a baby does not have many memories about the objects it experiences, there is a theory that consciousness expands slowly as the child gains more memories and thus can make more associations about the objects it perceives. On a physical level, although the brain is rich in connections between the cells, these have not yet been organized by experience.

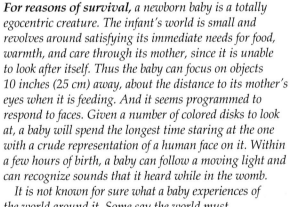

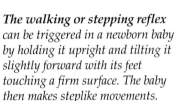

The Moro reflex can be seen when a baby is held flat on its back. If its head is allowed to drop slightly but suddenly, the baby will fling both of its arms out with the hands open and then gradually bend the arms back toward the body. This reflex usually fades by about four months. The Moro, stepping, rooting, sucking, and grasp reflexes are only a few of the so-called primitive reflexes that are present at or shortly after birth and that have faded by six months or so.

See also

FAR HORIZONS
▶ Having ideas
140/141

▶ Learning to be human
144/145

▶ Milestones of the mind
146/147

BUILDING THE BRAIN
▶ The brain plan
68/69

▶ The developing brain
70/71

▶ The cell factory
72/73

▶ Linking up
74/75

INPUTS AND OUTPUTS
▶ Survival sense
122/123

STATES OF MIND
▶ Being aware
158/159

▶ The resting mind?
160/161

▶ Drugs of abuse
178/179

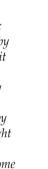

Learning to be human

The process of development is not confined to physical and intellectual growth; we develop emotionally, too.

A baby's relationship with its primary care-giver (usually the mother) has to provide the child with what it wants – food, warmth, and devoted care – so that it can develop properly physically and mentally. To achieve this, it is vital for the infant to establish a relationship with an adult that is close and rewarding. This usually happens right after birth, and some believe that close skin contact during these hours is vital for mother and baby to bond. Such a relationship seems so important that babies who receive no one-to-one bonding and love – such as those in some orphanages – have been known to die, although their physical needs were taken care of. From an evolutionary perspective, the apparent instinct for closeness has developed because it is vital to a baby's survival.

Babies are thus emotional creatures. Many mothers believe they can detect in their expressions a wide range of emotions, including interest, surprise, joy, anger, and fear, within a month or so of birth. By three months, other observers can reliably see them, too, and at that age babies can recognize emotions in other people, preferring a familiar or happy face to one showing another expression, for instance. By 18 months, infants can even fake emotions – acting peeved or hurt – to get what they want.

The development of interpersonal relationships starts from the earliest time and by childhood is fairly far advanced. The chances are that many of a person's character traits will also be established by this time. Play with other children seems to be important in helping children learn something of the social skills and emotional insights that are such an essential part of being human.

CUT OFF FROM THE WORLD

Children suffering from autism – a disorder that affects 4 in 10,000 – develop their motor skills very slowly, but more distressingly problematic, they seem unable to show any affection or to form emotional connections. Instead, they focus all their attention on objects and sometimes develop skills such as drawing to a high level. This image was created by an extraordinarily gifted child, Stephen Wiltshire, who has been diagnosed as autistic. Most autistic children hardly recognize parents or playmates and do not respond to them. They never look people in the eye and either ignore questions or reply by repeating them. However, they dislike change and may throw temper tantrums if the smallest detail of a daily routine alters. Until 20 years ago, mothers were blamed for creating autistic children by a cold and detached style of upbringing. But now it is accepted that the disorder is almost entirely genetic.

IMPRINTED ATTACHMENT

The fact that baby chicks will follow almost any moving object – another chicken, a duck, or a human – was first noted over 120 years ago, but it was animal behaviorist Konrad Lorenz who termed this instinct imprinting. Experimenting with baby geese, he showed that imprinting is instinctive, that it happens during a critical period after hatching, and that it cannot be reversed. In the wild, it is useful because the object the chick is most likely to follow is its mother. The strong bond developed between a human baby and parent is also thought to be instinctive and is obviously useful for the baby's wellbeing.

After establishing a strong early connection, babies often go through a phase, at around eight or nine months, of being frightened by strangers and fearful of separation – they object much more when left by their mother. This also seems like a useful adaptation for our hunter-gatherer ancestors to have developed in a world where sudden change or the unknown often meant danger – especially for a helpless infant. By 14 months, however, infants are eagerly exploring their world, occasionally coming back to a secure base for reassurance. The early years, when an infant is learning so rapidly, are just as important to its emotional wellbeing as they are in other areas of mental life, because patterns of emotional attachment formed in childhood can be repeated in adult relations.

Babies who enjoy a relationship involving high levels of love and care, especially physical contact and responsiveness to their needs, are more likely to develop a close attachment to their parents. A secure early attachment also tends to result in warm and loving adult relationships. Behavior benefits, too, so while such children enjoy being close to their parents, they are also happy to explore.

Children whose parents are withdrawn or distant, however, tend to stay detached from their parents and later keep partners at arm's length, too. Most complex are those whose parents alternated intense closeness with hostility or rejection. They become both desperate for a close relationship – thus they often make a great initial show of romance – and terrified the moment it seems likely.

See also

FAR HORIZONS
The evolving mind
126/127

Meeting of minds
134/135

Word power
136/137

First thoughts
142/143

Milestones of the mind
146/147

Male and female
148/149

SURVEYING THE MIND
The working mind
16/17

BUILDING THE BRAIN
Linking up
74/75

INPUTS AND OUTPUTS
Survival sense
122/123

STATES OF MIND
Emotional states
154/155

The unique self
156/157

The moral mind
182/183

Milestones of the mind

Slowly but surely the human mind develops until it is capable of phenomenal feats of logic and reasoning.

Because humans are such social creatures, we need a childhood that is about twice the length of that of any other animal if we are to absorb all the complex physical, social, and language skills we need to get ahead as adults. This long-drawn-out learning process can be affected by many different factors, from genetic makeup to the amount of love and stimulation we receive. But whatever happens, we are all preprogrammed to go through several definite stages.

For instance, we have baby teeth before adult ones, we crawl before we can walk, we understand stories before we can follow abstract ideas. The first big mental, as opposed to physical, leap comes at around 10 months, when babies first recognize themselves in a mirror. This is something that only a few species can do, and it marks the beginning of a sense of self-awareness, which later gives rise to a host of peculiarly human questions such as "Who am I?" and "Where have I come from?"

By the age of 18 months we are beginning to use symbols, and by two we have laid the ground for a number of crucial adult skills. We have arrived at half our adult height – if we continued to grow at the same rate, we would be about 11½ feet (3.5 m) tall by the time we reached adolescence. We have a vocabulary of about 200 words, which we use in two-word combinations. We no longer think of ourselves as the center of the world, we ask questions, we understand stories, we set goals and achieve them, we pretend, and we have formed attachments to people.

The next major milestone of change comes at puberty, when we have another growth spurt. This one can be accompanied by emotional problems, since it is combined with maximum self-consciousness; and growth is often uneven, leading to feelings of being gangly and awkward just at a time when attractiveness seems so important. Hormones set off the development of secondary sexual characteristics, and we become better at abstract thought and more responsive to ideals. And, as if the physical upheavals were not enough, we start to question everything.

Our mental abilities are thought to develop in definite stages, a theory most famously put forward this century by the Swiss psychologist Jean Piaget. He termed the first of these – from birth to about two – the sensorimotor period, when the child's senses (sensori) and muscles (motor) develop. Between two and seven is the preoperational stage, when a child uses symbols, but cannot yet manage abstract ideas. A key development of this stage is the realization that objects continue to exist, even if they are out of sight. *From 7 to 11, a child is deemed to be concrete operational, capable of logical thought, but only when it is tied to concrete reality. The formal operational stage, from 11 up, is when a child learns to make up classifications and build hypotheses.*

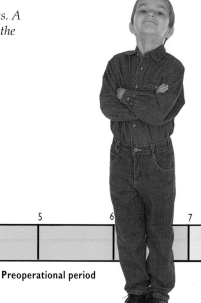

Sensorimotor period

Preoperational period

A famous experiment was used by Piaget to show that preoperational children could not see that the amount of liquid in a tall thin container was the same as in a short fat one. Since then, some have questioned his findings, but certainly by seven, children can realize that superficial changes to an object do not change its basic qualities.

See also

FAR HORIZONS
Word power
136/137

Having ideas
140/141

First thoughts
142/143

Learning to
be human
144/145

Male and female
148/149

**BUILDING
THE BRAIN**
The brain plan
68/69

The developing
brain
70/71

The cell factory
72/73

Linking up
74/75

**INPUTS AND
OUTPUTS**
Active vision
94/95

The mind's ear
98/99

**STATES OF
MIND**
Emotional
states
154/155

The moral mind
182/183

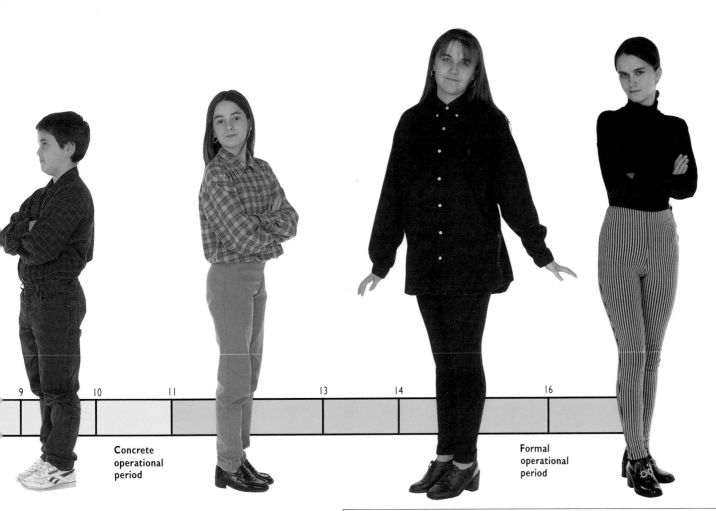

9   10   11   13   14   16

Concrete
operational
period

Formal
operational
period

Between birth and two, we develop from being able to recognize a face and remember tunes that we have heard in the womb to being able to walk, talk, and use a limited number of symbols.

By seven, we are less dominated by the senses and are developing the ability to classify things, for example into groups such as animals, red objects, things made of wood, and the like. At this age, we start to reason abstractly; have a grip on numbers, weights, and hierarchies; and are more ordered and focused.

Between about 7 and 11, we learn the basis of adult thought and reasoning, although we are limited to the concrete – reality as we have experienced it personally.

From 11 up, we make the jump from concrete thought to formal operational, or abstract, thought and deduction. We are now improving our skills in logic, making up theories about the world and trying out alternatives – a process that continues throughout adulthood.

CHILDHOOD GENIUS

Infant prodigy is an enduring subject of fascination, with some parents hoping to be able to push their children forward and give them a better start in life. There are many examples, from children who go to college at the age of only 11 or 12 to icons of the past like Wolfgang Amadeus Mozart (1756–1791), who could play the harpsichord when he was just three. A year later, he was playing

short pieces, and at five he was composing – an age at which any other parent would be proud if a child could pick out the tune of a nursery song. By the time he was 10, he had played at court in Munich and had been on a tour, performing at numerous cities in Europe. But can such genius be created? Probably not, nor can it be properly measured. The best a parent can do is to foster such gifts as their children display in whichever way seems appropriate to the child.

Male and female

Is there more to the difference between men and women than obvious physical characteristics and capabilities?

On average, men are 7 percent taller, 30 percent stronger, and 20 percent heavier than women; they also have more body and facial hair. Females have more developed breasts and tend to lay down fat on their thighs and buttocks, while weight in men gravitates to the stomach. Men are, generally, bigger and stronger, but women are designed for childbearing.

Psychological differences are less clearly defined, and more controversial. Men are perceived as more aggressive and competitive and less nurturing and emotionally responsive. And some researchers suggest that moral reasoning develops differently in the two sexes, with men tending to focus on abstract, rational principles such as justice and rights, while women see morality more as a matter of caring and compassion.

In terms of abilities, women on average talk more fluently, while men are better at tasks involving spatial ability, such as reading maps. And there is some evidence that male and female brains work differently. One recent study found that when asked to judge whether strings of nonsense words rhymed, men used only the left side of the brain, while more than half the women used both sides together. In another study, when people were asked to think of nothing, men's brains were active in the more primitive part that controls physical activity, while women's were active in the regions that deal with symbols and emotions.

There are great differences between male and female *in many species. The male golden toad* Bufo periglenes, *found in Costa Rica, is brightly colored and, at 1 inch (2.5 cm) long, much smaller than the mottled female. He clings to the female's back to fertilize her eggs as she lays them. There can also be differences in the roles of male and female. For instance, the male lion guards his pride against other males while the females hunt, and the male emperor penguin incubates the egg, but the female returns to feed the chick.*

The testosterone surge *a male fetus receives in the womb prepares it for later male development. There is also a rise in levels of the hormone in the first year after birth and a huge one at puberty (between 10 and 16), which produces the secondary sexual characteristics of mature genitalia, deeper voice, and a beard, as well as a growth spurt. After a peak in the early 20s, levels of testosterone fall slowly throughout a man's life.*

Female fetuses *develop ovaries, which secrete the hormone estrogen, by eight to nine weeks. An increase in estrogen levels at puberty (around age 12) produces a growth spurt, breasts, and mature genitals. This overall rise is subject to fluctuation during a woman's menstrual cycle. And after the menopause in middle age, estrogen production diminishes rapidly until almost none is secreted.*

DEFAULTING TO THE BASIC BLUEPRINT

A girl who reached the usual age of puberty, but still had "infantile" external genitalia and was short for her age, complained to her doctor that she was not developing properly. The doctor ordered tests, which revealed that she had only one sex chromosome – an X – a condition known as Turner's Syndrome.

At fertilization, each parent supplies one half of each of the 23 pairs of chromosomes. One of these pairs decides sex: two Xs give a girl; an X and a Y make a boy. A Y chromosome, which can come only from the father, imposes maleness on the basic human blueprint, which is female. The girl was given hormone treatment and developed secondary sexual characteristics and monthly vaginal bleeding, but was infertile since two Xs are needed for normal ovary development.

Hormone levels

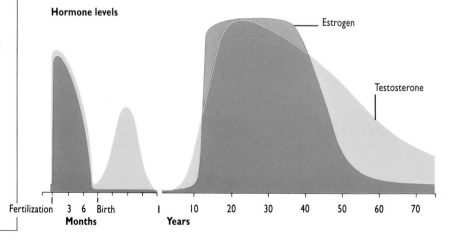

Estrogen

Testosterone

Fertilization | 3 6 Birth | 10 20 30 40 50 60 70
Months | **Years**

The sex of a child is genetically determined, but where, if at all, do children learn to be different and to conform to the behavior that is deemed to be typical of their sex? Some theories state that this learning comes from the culture and from the expectations that exist within it. Others argue that the learning of roles and differences in behavior come from identifying with an adult of the same sex. Still others suggest that the way children and then adults behave is genetically predetermined: the various genes and all that they are responsible for in terms of hormones and other basic differences cause not only the physical differences, but also any psychological ones. It seems that children have some predetermined differences but that they also probably learn from the society around them and attempt to live up to its expectations, being rewarded when they do and discouraged when they do not.

Whatever the case, traditionally, the ways girls and boys are treated start to differ almost immediately. In some cultures, this means a boy is dressed in blue, a girl in pink. And although parents may try to avoid sexual stereotyping, boys often choose to play with a toy car rather than a doll, and girls may prefer pretty feminine dresses to jeans. As they grow older, boys will often, though not invariably, choose competitive games while girls cooperate more in their play.

See also

FAR HORIZONS
▶ Parallel minds
138/139

▶ First thoughts
142/143

▶ Learning to
be human
144/145

▶ Milestones of
the mind
146/147

**SURVEYING
THE MIND**
▶ Brain maps
30/37

**BUILDING
THE BRAIN**
▶ The brain plan
68/69

**INPUTS AND
OUTPUTS**
▶ Dealing with
drives
118/119

▶ Rhythms of
the mind
120/121

**STATES OF
MIND**
▶ Emotional
states
154/155

▶ The moral mind
182/183

Thinking together

Our ability to think and share our thoughts and ideas has played a vital part in the human success story.

Much of psychology is involved with the study of individuals – how they develop, what their personality is like, how their brain works. But it is essential not to lose sight of the fact that we are intensely social animals who can exist only in a complex web of relationships. Society strongly shapes our behavior and gives meaning to our lives. Indeed, compared to other animals, human children need a firm social structure over a long period before they become adult.

Among early humans, the value of improved social skills was probably a powerful factor driving the development of both self-consciousness and language. Groups in which people could communicate often complex ideas – only truly possible with speech – would have been at a great advantage. The success of *Homo erectus*'s migration out of Africa about a million years ago, for instance, can be attributed to improved group organization, and the Neanderthals' failure to compete with modern humans some 950,000 years later is thought to be linked to their failure to form large groups.

Modern humans are born with a powerful urge to communicate with others: a newborn baby not only responds very soon to faces, but also tries to gain a response. People will do almost anything to win approval and acceptance by the group, and many psychological experiments have shown that they can be made to alter their behavior drastically

The pyramids of ancient Egypt are seen by many to epitomize the success of group effort. In the days before mechanization, these vast monuments to a culture could have been achieved only by the intense and dedicated labor of thousands of people working together.

Something of the same group effort is shown today in industry where, for instance, teams of workers, each one contributing his or her own particular skill, produce the thousands of automobiles that roll off the production line every year.

An earthquake such as the one that hit Mexico City in 1985 is just one of the numerous situations in which it is essential for us to work together if we are to make any major impact. People facing such a disaster will rapidly identify with those who have experienced the greatest suffering and will become selfless members of a group toiling to rescue victims at considerable risk to their own lives.

In such a situation, it will soon become necessary for a leader of some sort to emerge, and experiments have shown that the army model of an autocratic leader and obedient followers may prove the most effective, since it results in the highest level of productivity by the group. Under other conditions, a more democratic approach may prove valuable, encouraging people to cooperate and share responsibility rather than blindly obeying orders or rules.

because they believe that a certain type of behavior is expected by the group. Indeed, experiments have shown that students are prepared to administer quite severe electric shocks to their fellows if enough pressure is exerted upon them to obey the mores of the social group and so to "belong."

One of the strongest sanctions that can be applied to anyone is expulsion from the group. Once it meant almost certain death; today, complete loners are more likely to have severe mental problems, while those who have lasting and close connections with others tend to be healthier and happier. An extreme example of the human need for togetherness can be seen in a common attitude to times of war, which are often fondly remembered – despite the hardship and suffering they entail – because of the sense of connection they create. And part of the enduring appeal of soap operas on radio and television is that they provide the illusion of being involved with a constant and familiar community.

In contrast with this is the powerful image of the lone genius in the arts or sciences, but even the most gifted base their work on what has gone before – there could, for instance, have been no Einstein without Newton. This forms the basis for our culture, for we are connected not only to the minds of those around us, but to the minds of the past as well.

See also

FAR HORIZONS
▶ The evolving mind 126/127

▶ Meeting of minds 134/135

▶ Word power 136/137

▶ Having ideas 140/141

SURVEYING THE MIND
▶ The working mind 16/17

INPUTS AND OUTPUTS
▶ Survival sense 122/123

ANIMAL SOCIETIES

Like humans, animals gain enormous benefits from living in societies. They can gang up on predators such as eagles or lions, or they can hunt in packs like wolves and bring down prey they could never manage to kill working alone. Some of the most complex societies are found among insects such as termites, ants, and bees. These societies, with their single leader or queen and rigid castes, including soldiers, drones, and workers, are sometimes held up as a mirror of certain human societies. The crucial difference, however, is in their lack of flexibility.

While humans do create rigid, centralized, totalitarian societies, we can also form liberal, democratic, and open ones. Individual ants or bees are not at all like the individuals in a human society: the ants have no choice but to follow the dictates of their genes as to how they should behave. We, however, can change, and if enough people want a different type of society, they can sometimes create it by concerted action – as with the fall of the Berlin Wall in 1989 – for the power and influence of the group will is immense.

The success of social insects such as ants is clearly evident in the Amazon basin, where although they make up only 3 percent of the insect species, they form 80 percent of the total number of insects found.

STATES OF MIND
▶ Talking cures 164/165

▶ The moral mind 182/183

States of Mind

Consciousness is the most amazing of the brain's qualities. Without it we would be little more than automatons, merely going through the motions of life. We would, for example, have no appreciation of such aspects of existence as the subtle interplay of our own emotions and the richness of relationships with other people, each with a unique personality.

Without it, we would not be able to question what we do and why we do it, or to ponder moral questions and debate issues such as the existence of free will.

But the price for possessing such a remarkable mind can be high. We all experience sadness and anxiety at times, but some people become overwhelmed by abnormal states of mind and become ill. Help is at hand, though, and most recover to savor again the experiences self-awareness brings.

Left (*clockwise from top*): *a tiny mind; object of fear; blot on the personality; a lemon squeezer made into a desirable object; a slice of sleep.* **This page** (*top*): *if birds were neurotransmitters;* (*left*) *food, drink, and depression.*

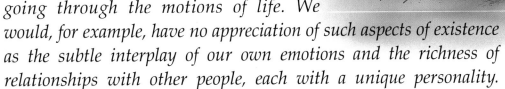

Emotional states

Whether you are aware of it or not, emotions suffuse everything you do, coloring every thought and action.

Happiness, rage, love, and hate are just some of the feelings that make us human and not machines. They seem illogical, but may be vital to our survival as a species. Love, for instance, may have evolved to guarantee we not only find a partner, but stay with him or her and make sure our offspring survive. Anger and fear may help us fight off or run away from danger.

People from different cultures label emotions in different ways, but happiness, fear, anger, sadness, and disgust are universal and are linked with facial or vocal expressions that are understood worldwide. Other emotions such as surprise or guilt – and the expressions associated with them – vary between cultures.

For each emotion, there are three elements. The first is how we feel subjectively – happy or sad, for example. The second is change in the body over which we have no conscious control, such as butterflies in the stomach or goose pimples. The third is the behavior we associate with a particular emotion – running away, laughing, and so on.

Different theories about emotion emphasize a different one of these elements. One view is that the feeling causes the change in the body. Another is that changes in the body create the feeling: when you tremble, the trembling makes you feel

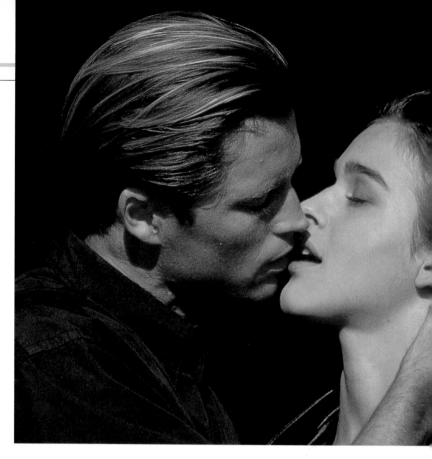

A lovers' kiss is the perfect expression of that most discussed of all emotions – love. Some psychologists say loving is just a more intense version of liking. Others say loving and liking are very different. Freud saw love as the socially acceptable face of the libido, or sex drive. It has also been suggested that there are two kinds of love: one which stems from our need for security and inevitably leads to disappointment because it depends on the other person to satisfy it; the other which is more settled, with both partners secure and balanced. Other theories of love center on the difference between brief all-absorbing passions and long-term affection, or on the varying balance between intimacy, passion, and commitment.

CURBING HIS PASSION

A few years ago, a man in California was arrested lying on a sidewalk trying to make love to it. When asked why, he said he found sidewalks irresistible. Cases like this are rare, but animal experiments in the 1930s showed how brain damage may alter emotional responses. Damage to the cortex, for instance, produced "sham rage" in cats, in which the cats showed all the signs of rage, yet purred and allowed themselves to be stroked. Damage to the amygdala of monkeys made them calm, but their sex drive became so active that they mounted anything they could and put all kinds of objects in their mouth, a change known as Klüver-Bucy syndrome after the scientist who carried out the experiment.

Attempts to connect particular emotions, such as aggression, completely to particular parts of the brain have failed, but the amygdala does seem to play an important role in emotions, linking the physical signs of emotion triggered by the hypothalamus with the mental experience of emotion in the cortex.

scared; when you smile, the smile makes you happy. Others think the feeling and the bodily changes are independent, since people who are paralyzed feel emotions just as intensely as anyone else. They believe an emotion starts when the stimulus that triggers it is processed in the brain by the thalamus. The thalamus then sends signals simultaneously to the cortex, where we become aware of the emotion, and to the hypothalamus, which sets off changes in the body via the nervous system and hormones. Another theory, that of cognitive labeling, suggests that the body becomes aroused before we are aware of a particular emotion – but the emotion we feel depends on what we think aroused us. In one experiment, people injected with epinephrine laughed more at a comedy than those without the injection. In another, men on a dangerous rope bridge found a woman interviewer far more sexy than did men on a solid wooden bridge.

See also

STATES OF MIND
▶ The unique self
156/157

▶ Being aware
158/159

▶ Abnormal states
162/163

▶ Feeling low
168/169

▶ Anxious states
172/173

▶ Motivation
180/181

▶ The moral mind
182/183

INPUTS AND OUTPUTS
▶ Keeping control
116/117

▶ Dealing with drives
118/119

▶ Survival sense
122/123

FAR HORIZONS
▶ Learning to be human
144/145

THE SPECTRUM OF EMOTIONS

Many psychologists believe emotion is bound up in a continuous loop of feedback between bodily changes and the feeling we are aware of. One way of looking at it may be as an interplay between arousal, drive, and association. Our bodies and brains become aroused by certain stimuli – a dangerous situation or the sight of an attractive person, for instance. This arousal may also be influenced by the second of these elements – basic physiological drives, such as the need for food, sex, or self-preservation. The third element is the whole range of mental associations that flood into the mind.

In the image below, a white rose is illuminated with red, blue, and green light. Like the subtle interplay of the lights on the rose, the shifting balance between the three elements of association, drive, and arousal determines the nature of the emotion we feel and the way our bodies respond.

Associations – essentially memories or ideas that spring to mind – can move through the mind at an extraordinary rate, adding texture and context to emotion. They can also alter arousal levels. Associations linked with seeing the rose may involve memories of the last time a rose played a part in a person's life. This could be a pleasant association, such as giving or receiving roses as a love token, or an unpleasant one, such as being pricked by a thorn on the stem. For some, the association may be sinister since in some cultures white flowers are unlucky and can be reminders of funerals or death.

Arousal level affects emotional response. In a sense, it acts like a volume control of emotion. If arousal level is high to begin with, the response, to the rose in this case, is strong; if arousal is low, the response is weak. Seeing the rose is unlikely to change your arousal level by itself, but it may alter it by evoking an association.

Drives have a constant input into emotional state. If you are hungry or thirsty, for example, the way you feel about anything is affected by whether it is likely to satisfy the drive or not. By itself, a rose in a vase will probably not have much impact on drives since it cannot ordinarily be eaten, drunk, or made love to. However, if an association of love comes to mind, the appropriate drive may well swing into action.

The unique self

Each of us has a different personality – our individual pattern of emotion, behavior, and perception.

If you say "I am not myself today," everyone knows what you mean. But it is an odd statement. Who are you then? A string of everyday words – such as kind, happy, open – might seem to describe your "personality," but even psychologists have difficulty in getting to grips with the concept. Although you might think that you know what type of personality you are and that your behavior is consistent with those traits, in some situations you may do something "out of character."

There are many theories as to what constitutes personality, how and whether it can be measured, and how it develops. Some see it as a result of learning throughout life; others relate it to the environment, society, or family; still others believe it is genetic or developed in early childhood. And personality is not set in stone – people can change, either as a result of altered circumstances or because they have made a conscious effort to change, perhaps with the help of psychotherapy.

There are also recognized personality disorders. One survey has concluded that nearly one in ten of the adult population in the United States may suffer from a disorder in which that person has traits that affect functioning – socially or at work – or cause mental anguish. A few people are even diagnosed as having multiple personality disorder, in which they have two or more distinct personalities between which they switch.

***Extroverts and introverts** – sociable partygoers and withdrawn stay-at-homes – are two of the broad categories of personality. Extroverts need excitement and stimulation, so they tend to be outgoing, enjoy driving fast, and like working with people. Introverts already have a busy inner life so they need little additional excitement.*

FOOLING ALL OF THE PEOPLE ALL OF THE TIME?

Astrology is one ancient system for describing personality which many people find to be accurate. However, this so-called accuracy may in fact be because we are not particularly adept at identifying our own personality, not because it is governed by our date of birth and the relative positions of the planets in the heavens at the time.

A classic psychology experiment involves giving a personality assessment to each member of a class and asking them to say how accurate it is. About 90 percent report that it is either a good or excellent reflection of their qualities. At the end of the test, the experimenter reveals that everyone received exactly the same profile.

The key to the assessment is that it is made up of carefully constructed sentences, known as Barnum statements, which almost everyone agrees with, but – and this is the clever part – most do not believe also apply to everyone else. An example is: "Normally you are quite outgoing, but underneath you are more shy." So the reason that so many people accept astrological descriptions of personality as correct is that most consist almost entirely of Barnum statements.

DESCRIBING PERSONALITY

There are 18,000 adjectives in the English language to describe people, and many of them fall into clusters. So someone who is kind is often also thought of as helpful, friendly, and thoughtful. Since the time of the Greeks, we have assumed that these clusters point to some underlying aspect of personality. But how many clusters are necessary to describe the basic personality types?

In the second century, the Greco-Roman physician Galen outlined three clusters: the cognitive, covering the intellect; the conative, referring to intuition; and the affective, for the emotions.

A medieval system, based on one put forward in the fifth century B.C. by the Greek physician Hippocrates, had the four humors. They were fluids in the body, the proportions of which gave rise to different personalities. Blood produced a happy and lively – or sanguine – character; phlegm was linked with calm, careful, "phlegmatic" people; black bile caused melancholy and pessimism; yellow bile led to a "choleric" nature involving hastiness and excitability.

Since then, despite many statistics and extensive personality testing, we have not progressed very far. Psychologists today talk about personality dimensions or traits. One model, now rarely used, says that 100 distinct traits need to be considered to describe personality adequately; Raymond Cattell, a leading personality theorist, developed a system using 16. At the other extreme is Hans Eysenck who believes that only three basic dimensions are necessary – extroversion–introversion, psychoticism, and neuroticism–stability.

Today the most generally accepted trait theory is that of the "big five," or the "five robust factors," which works with the five personality traits that recur most often in studies and theories. The most commonly occurring five characteristics can be summed up in the acronym OCEAN: Openness, covering such aspects as intelligence, the imagination, and esthetic sensitivity; Conscientiousness; Extroversion; Agreeableness, meaning someone who is good-natured, friendly, and easy to get along with; and Neuroticism.

The Rorschach inkblot test is one of the better-known methods used for assessing personality, but it is not often applied outside of psychoanalytic circles. It was devised in 1921 by Hermann Rorschach and consists of 10 symmetrical inkblot designs, each one printed on a separate card. Five of the designs are in black, white, and shades of grays, five are in color, and all are non-representational.

When subjects describe what they see, aspects of their psychological make-up can be revealed. Most people will project themselves into the designs, each person interpreting an image in a different way – and thus helping the psychologist to establish elements of that person's personality or a particular mental condition.

For instance, with the colored blots, such as this one – the only one to be fully colored – the subject's response to the colors is believed to reflect his or her emotional life. And it is not merely detailed descriptions of color that are important, failure to mention the colors at all can be equally significant.

See also

STATES OF MIND
▶ Emotional states 154/155

▶ Being aware 158/159

▶ Abnormal states 162/163

▶ The moral mind 182/183

▶ Individual minds 184/185

SURV, THE M
▶ The wo mind 16/17

▶ Inside the mind 20/21

BUILDING THE BRAIN
▶ Remaking the mind 62/63

▶ Linking up 74/75

FAR HORIZONS
▶ Having ideas 140/141

▶ Thinking together 150/151

Being aware

The fact that you know you are reading this illustrates the brain's most amazing quality, self-awareness.

We all know what consciousness is like from the inside. It is where we think and feel and remember. But scientists must analyze it from the outside, and so come up against the problem that there is no way to measure it, since it exists in a private world. Researchers studying such subjects as the workings of muscles can all look at the same thing, but consciousness belongs to individuals, and the only person who can know what your consciousness is like is you.

Nonetheless, consciousness is obviously dependent on the brain. Drugs affect it, some even sending it crazy or knocking it out altogether. The key question is whether it can be explained entirely by the firing of brain cells. Some researchers believe that it can and that once enough information is gathered, it will be possible to say that activity in a particular combination of neurons means that the subject is having a particular sensation or thought. Others claim that the very nature of consciousness means that it cannot be explained in such terms. They focus on the idea that a process of natural selection takes place among brain cells from the moment of conception, driven by the different experiences that each individual has. The result is that the precise structure of each person's brain is unique. Another argument is that some brain processes take place at a quantum level, which means they cannot be described only in terms of chemical and electrical processes.

At the heart of the debate is the puzzle of how matter – molecules – and electricity can produce something that is aware of itself. We do not just hear and respond, as a machine does, we hear and know what it is like to hear. Some claim this awareness is just a by-product of the brain's complexity – brains produce minds like clouds produce rain – others say there is a gap there that science as we know it now cannot bridge.

Dressing up is one of the ways children have of playing with newly emerging ideas of self-consciousness. To pretend you are someone else, you must first of all be aware of yourself as a separate individual. Only a few monkey species show signs of being able to do this, but by the age of four, children seem able to grasp that other people have a conscious internal world as well.

Consciousness is remarkably difficult to define. Some describe it as self-awareness – that quality that differentiates humans from animals. Others use the term "soul"; still others refer to it as the mind. Are these terms all interchangeable, depending on a person's point of view, or do they refer to different aspects of the mind? A simple distinction is that a soul is something enduring, believed by many to last beyond the life span of the body or physical brain. The mind is limited by the duration of a particular life. Consciousness is what we have during waking hours, while awareness is our grasp of the here and now.

Soul is forever  **Mind** is for life   **Consciousness** is for today   **Awareness** is for now

DEAD OR ALIVE?

A young man who suffered severe head injuries in a car wreck was in a coma – a state where the brain does not work at all – for some weeks. He then regained some functions: his sleep–wake cycle was restored, and he was able to regulate his temperature, breathe and maintain his heartbeat, and to swallow and digest food. He also withdrew from pain, could blink, and his pupils contracted if a light was shone on them. He could also make some noises, such as moans, and occasionally seemed to smile. But none of his behavior – apart from the reflexes – bore any relation to what was going on in the outside world.

He was eventually diagnosed as being in a persistent vegetative state (PVS), a condition in which, despite signs to the contrary, patients are not aware, not conscious, and are said to have suffered cognitive death, (cessation of thinking). His brain damage was in the higher regions, the cortex, while his signs of life were generated by the lower parts of the brain where most autoregulatory and reflex responses are dealt with. After a few years with no signs of improvement, his family took medical advice and requested that feeding be withdrawn and the man died.

Just as the flavor of a stew is more than the sum of its ingredients, our mental image of an object is a composite structure that emerges when various different elements come together. The vision centers in our brain break down what we see into these elements – straight or curved lines, edges, and so on – and then combine them to produce faces, trees, chairs. But we cannot understand an object by learning more and more about curves and lines.

Some neuroscientists believe that consciousness emerges as different groups of neurons – dealing with vision, memory, or touch – combine and recombine. A sudden sensory input – a car braking hard, for instance – can trigger a completely different combination of neuron groups from the conscious state produced by reading or listening to music.

In this view, just as there is no single center for vision or language, there is no seat of consciousness, no internal theater where consciousness is a permanent spectator. Instead, what we experience as consciousness is this constant procession of waxing and waning neuronal groupings.

See also

STATES OF MIND
▶ Emotional states
154/155

▶ The resting mind?
160/161

▶ Individual minds
184/185

SURVEYING THE MIND
▶ Discovering the brain
14/15

▶ The working mind
16/17

▶ Inside the mind
20/21

▶ Electronic minds
22/23

INPUTS AND OUTPUTS
▶ Levels of seeing
92/93

▶ Active vision
94/95

▶ The big picture
114/115

FAR HORIZONS
▶ The evolving mind
126/127

▶ First thoughts
142/143

The resting mind?

During sleep, a time when body and mind are restored, the brain regularly switches on and dreams.

When we fall asleep, we go through four stages of deepening sleep measured by changes to EEG (electroencephalogram) recordings which measure the brain's electrical activity. Moving through the stages, the brain waves get progressively slower. This lasts about 80 minutes, and we then return to a state in which the brain waves look like those produced when we are awake. This state is known as REM (rapid eye movement) sleep, and this is the time when we dream. These eye movements – in which the eyes flick back and forth – are the only visible sign of a number of dramatic changes. Brain cells at the top of the spine send out signals that prevent any movement so we are virtually paralyzed, except for our eyeballs. Also activated is our balance system in the ear and the visual part of the cortex. After about 5 to 15 minutes of REM sleep, we sink back down into deep sleep once more. We go through this cycle four or five times a night.

Throughout history, people have asked why we dream and what dreams mean. Until about 100 years ago, their source was thought to be partly supernatural, perhaps involving messages from the gods. Then pioneer psychoanalyst Sigmund Freud replaced the supernatural with our own unconscious when he stated that dreams are the expression of unfulfilled desires, the more shocking of which pass through an internal "censor" that replaces them with symbols which psychoanalysts attempt to interpret. Another theory is that we dream in order to sort out memories, either adding them to the memory store or throwing away unwanted information.

More recently, it has been suggested that dreams are attempts by the brain to make sense of stray thoughts. During the day, our cortex interprets the information that floods in from the senses to create a picture of the world. At night, with nothing coming in, it tries to interpret the weak internal scraps of thought and link them with stored memories.

At 26 weeks, a fetus is in REM (rapid eye movement) sleep 24 hours a day. In adults this is the stage at which we dream. What can a child in the womb be dreaming about? One theory is that REM sleep at this time could be nature's way of giving some of the brain's systems a trial run before the baby emerges into the real world.

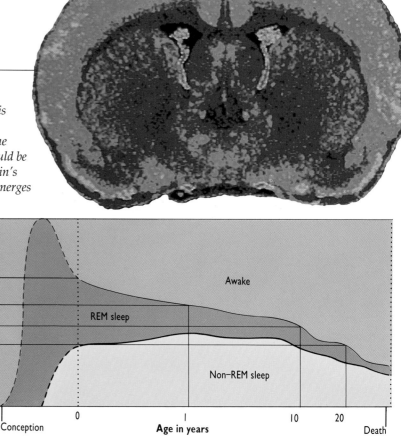

Awake

The proportion of time we spend awake, asleep, or dreaming (REM sleep) changes during our lives (right). By the time a baby is born, it still spends 16 hours a day asleep, half of those in REM sleep. As we get older, we sleep less and spend less time dreaming.

Hours

Awake

REM sleep

Non–REM sleep

Conception   Age in years   Death

THE IMPORTANCE OF SLEEP AND DREAMS

☤ In 1959 New York disc jockey Peter Tripp stayed awake for 200 hours to raise money for charity. He was supported by doctors and psychologists who tested his mental functioning. But his prolonged period of sleep deprivation had unexpected and unpleasant results. After about 50 hours, he started having mild hallucinations, seeing cobwebs in his shoes when there were none there and thinking that specks of dirt were bugs; by 100 hours, he found simple mental tests unbearable. At 110 hours, he became delirious and saw a doctor's tweed suit as a tangle of furry worms; at 120 he needed a stimulant to keep him awake. After 150 hours, he was disoriented, not knowing who or where he was, and he became paranoid – he backed against a wall, letting no one pass behind him; by 200 hours, his hallucinations had taken a sinister turn, and he thought a doctor trying to examine him was an undertaker come to bury him. When it was over, he slept for 13 hours, after which his paranoia and hallucinations had gone.

Oddly, when it was time for his regular broadcasts – between 5 and 8 p.m. – he was able to perform almost normally. He could carry out relatively complex tasks such as playing records and was able to talk in a fairly normal fashion to his listeners. Other, more formal, sleep-deprivation studies have had similar results. For instance, Tripp's ability to rouse himself for his broadcasts bears out the fact that subjects of experiments who have been seriously deprived of sleep can revive themselves to perform brief tasks.

REM sleep (dreaming), in particular, seems to be vital to mental health. If people are deprived only of REM sleep, they will have psychological symptoms; when allowed to sleep normally again, they make up for it by spending more time proportionally in REM sleep.

REM sleep

Non–REM sleep

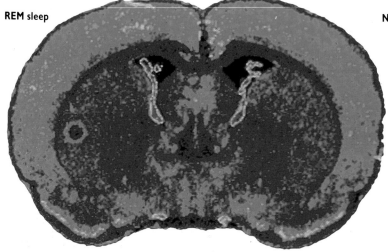

Maintenance and repair work takes place throughout the body while we sleep. Evidence for this in the human brain comes indirectly from another mammal, the rat.

Images taken of slices of the brains of rats (**above**) reveal that protein synthesis – which is necessary for reconstructing cells – increases in the brain during non-dreaming, or non-REM, sleep compared to wakefulness or dreaming (REM sleep). In humans,

during dreaming the amount of the brain chemical norepinephrine is also sharply reduced. Since norepinephrine seems to be involved in the process of organizing information into a coherent form, its low level during dreaming could explain why dreams appear to disregard logic and seemingly impossible things can happen.

Elsewhere in the human body, other physical changes occur during sleep. For instance, blood pressure

drops, and breathing and heart rate slow down, but more growth and sex hormones are released.

Nearly all animals, including humans, have some sort of rhythm of rest and alertness, although the reasons for this are not entirely clear. It is thought that in addition to allowing the body to repair and maintain itself, sleep could be a survival device – a time to conserve energy during periods of inactivity.

Protein synthesis

Low

High

See also

STATES OF MIND
▶ Being aware 158/159

▶ Talking cures 164/165

SURVEYING THE MIND
▶ The working mind 16/17

▶ Probing the mind 26/27

▶ Brain maps 30/37

BUILDING THE BRAIN
▶ Building blocks 40/41

INPUTS AND OUTPUTS
▶ Keeping control 116/117

▶ Rhythms of the mind 120/121

FAR HORIZONS
▶ The infinite store? 130/131

▶ First thoughts 142/143

Talking cures

Physical cures for mental problems mostly treat only symptoms, but some talking cures can deal with causes.

When a person is physically ill, there is often just one recognized treatment. But when the problem is mental, this is far from so; there are almost as many different therapies as there are psychiatrists. Some approaches are physical, such as using drugs; others involve talking. Yet it would be a mistake to assume they are mutually exclusive. Many psychiatrists will treat a depressed patient, for instance, both by talking about the problem and with antidepressant drugs.

The talking cures come under the umbrella heading of psychotherapy and fall into four main areas: psychoanalytic, humanistic, behavioral, and cognitive. By far the best known is psychoanalysis, developed by the legendary Sigmund Freud. This involves skilled probing into the patient's past to reveal the origins of unconscious repressed feelings, thoughts, and

Fear of flying, however irrational, is something that can be crippling for sufferers. One treatment that can be used to help such phobias is behavior therapy, which essentially deals with the abnormal behavior itself and not necessarily any deeper cause. There are various approaches. Systematic desensitization teaches the patient to relax as he or she is exposed to anxiety-producing situations in a series of graduated steps, starting with something relatively minor and, when relaxation can be achieved with that, moving up to something stronger. This aims to replace one response to a particular stimulus with another. Implosion therapy, a similar but more extreme approach, involves surrounding the patient with the source of fear – not building up to it gently – on the basis that the phobia will soon subside. Operant conditioning works by giving rewards for changed behavior.

Group therapy is one alternative to the more common pattern of one-to-one psychotherapy and has been popular since World War II, particularly with the increasing use of psychotherapists amid restricted resources. For many problems, it is not only less expensive, but some therapists also believe that it can be as useful as individual treatment.

In a typical group therapy session, the patients and therapist discuss the problems of one or more of the members. There are a number of reasons why this can be helpful. Perhaps the most obvious is if the group is set up for people who share a common problem, alcoholism or bereavement for instance. They can then share their experiences about that problem and come to a deeper understanding of it – and their reactions to it.

It has also been argued that other aspects specific to group therapy are valuable. The strength gained from being a member of the group is one of them – patients feel supported and so are more able to take risks and cope with criticism. Another is the sense of self-worth gained by helping other people. Members often also begin to understand themselves better – partly through seeing similar concerns and reactions in others – and learn to express themselves better.

See also

STATES OF MIND
▶ Emotional states 154/155

▶ Abnormal states 162/163

▶ Physical cures 166/167

▶ Feeling low 168/169

▶ The mind adrift 170/171

▶ Anxious states 172/173

▶ Fears and fixations 174/175

SURVEYING THE MIND
▶ The working mind 16/17

FAR HORIZONS
▶ Active memory 132/133

▶ Word power 136/137

▶ Learning to be human 144/145

motives. Humanistic therapists, by contrast, see the problem as a problem with how they perceive themselves, and they refer to the people they see as clients, not patients. They will not delve for unconscious motives, but simply provide the support clients need to help them find their own solution.

Psychoanalysis, including the schools of psychotherapy derived from it, and humanistic therapy both tend to be long-term treatments – psychoanalysis can take years. Behavior therapy, however, tends to be quick and direct, treating only the immediate symptoms. If a person is anxious, a behaviorist will not look for an underlying cause, but will simply reduce the anxiety with appropriate stimuli. Cognitive therapy seeks to condition (teach), but here the aim is to provide the patient with a better way of thinking about a problem, since the theory is that our perception of something – distorted in the case of someone with mental illness – governs how we respond to it.

Besides these four mainstream approaches, there are myriad alternatives. Gestalt therapists, for instance, focus on developing awareness, arguing that problems arise when we are unaware of our needs and leave them unfulfilled. Advocates of transactional analysis (TA) believe it is useful for everyone, not just those with "problems." TA aims to make us aware of the way we and those around us behave, and of the games we all play, so that we can direct our behavior accordingly.

DIGGING INTO THE PAST

Sigmund Freud was the first to suggest that the causes of psychological problems could be uncovered by delving into a patient's past. He did this by encouraging his patients to lie down on a couch and relax, and then used such tools as free association and the interpretation of dreams. With free association, the fully relaxed patient says whatever comes to mind, without censoring or editing it; the analyst simply acts as a guide. The repressed memories causing the conflict and undermining the patient's ego only emerge slowly because people subconsciously offer resistances – blocks that save them from confronting painful memories. Whenever the painful topic looms, the patient may fall silent or make jokes. After a while, the analyst will try to help the patient interpret these memories. The patient may strongly resist the analyst's view at first, but this only confirms the interpretation, because strong resistance must conceal painful conflicts.

Psychoanalysts aim to remain emotionally detached from their patients, since by doing so they become a blank screen, encouraging patients to shift feelings and thoughts they have toward others – both positive and negative – onto the analyst. This process, known as transference, helps bring out the suppressed conflicts.

Physical cures

There are some powerful remedies available for relief of mental symptoms.

Much mainstream psychiatric treatment relies upon physical rather than psychological methods. Physical treatments fall into three groups: psychosurgery, electroconvulsive therapy (ECT), and drug therapy. The first two are controversial and not common, whereas the use of drugs is relatively uncontroversial and widespread. Drugs have, in fact, revolutionized treatment of mental illness since they were first widely used in the 1950s. Now millions of patients worldwide benefit from drug therapy.

But psychiatrists who rely too heavily on drugs have been criticized because of the side effects of some and because of the dependency others encourage. Other doubts arise since many of the major drugs were discovered by accident, and it is not known why their action at the cellular level is therapeutic. It is often possible to see how they operate on certain neurotransmitters in the brain – but no one knows why this has particular psychological effects. And no drug actually cures a problem, it simply relieves symptoms.

Four main groups of drugs are used to treat mental illness: antipsychotics; antidepressants; antianxiety drugs; and those for manic-depression. Antipsychotics commonly used to treat schizophrenia are phenothiazines such as chlorpromazine (Thorazine). They are thought to work by blocking the receptors for the neurotransmitter dopamine in the brain – particularly in the limbic system, which is deeply involved in our emotions, and the hypothalamus, which is involved in our basic drives such as sex and food. By blocking the nerve impulses transmitted by dopamine, they dampen down emotional response and interrupt basic drives, reducing symptoms.

Antidepressants were discovered by accident in 1952, when tuberculosis patients receiving the drug iproniazid started getting "high." This, too, works

Psychotropic (mind-affecting) drugs almost all have their action at the synapse, the gap between nerve cells (neurons). They modify how one neuron communicates across the gap with another neuron. Some drugs, for instance, alter the amount of neurotransmitters (chemicals that carry messages across the synapse) made in the sending neuron. With more available for release, the chances are that communication will be stronger. In the same way, if carrier pigeons breed, there will be more birds to carry messages.

About 50,000 people – the number of M's in the Manhattan telephone directory – had prefrontal lobotomy operations in the '40s, '50s, and '60s, in which the frontal lobes of the brain are cut away. Usually performed on schizophrenics, the operation often left patients much calmer – but only because they were virtual vegetables. Psychosurgery is performed very infrequently today.

A SHOCKING BUSINESS

All attempts at alleviating the severe depression of a patient in a psychiatric hospital failed. After some months, doctors resorted to electroconvulsive therapy (ECT), which consisted of six treatments over two weeks. At the end of that time, the patient's spirits lifted and eventually he recovered.

The idea behind ECT evolved when it was noticed that epileptics who suffer convulsions are rarely schizophrenic. This led people to attempt to cure mentally disturbed patients by inducing convulsions. Eventually, a jolt of electricity to the brain, delivered via electrodes attached to the head, was chosen as the means of setting off the convulsions. The current causes a brain seizure lasting a few minutes, in which neurons all over the brain fire out of control. In the early days, people were often hurt as they thrashed about after the shock. In modern treatments, a muscle relaxant and anesthetic are given to prevent damage and to reduce anxiety.

ECT is used only for severe depression when all else fails; it often seems to work. No one is quite sure why it works, although one theory is that it increases neurotransmitter activity and corrects a deficit in norepinephrine levels and perhaps also in serotonin. Some people experience memory loss and confusion after ECT, but this is usually temporary.

But its use is controversial, and clinicians have widely differing opinions about it. It is viewed by some as an extreme and frightening measure and this, together with associated memory problems and uncertainty about how it works, has led to a marked decline in its use since its 1950s heyday.

STUMBLING ON A TREATMENT

In 1949 Australian psychiatrist John Cade (**right**) was carrying out experiments on guinea pigs to test a theory that high levels of uric acid in the body caused mania. To facilitate absorption of the uric acid by the guinea pigs, he mixed it with lithium. But when he injected the mix, he noticed that the animals grew calmer, not more lively as he had predicted. He began to suspect that the calming might be due to the lithium, and when he gave it to a group of manic patients, he found that it calmed them down and

regularized their mood. At the time, lithium had just been banned in the U.S. because it had caused several deaths when used as a substitute for common salt. It was not until 1970 that it was approved for use in treating bipolar disorders (manic-depression).

Lithium was a breakthrough and is effective in reducing both the up and the down mood swings of bipolar disorder. Its use has to be carefully monitored, however, since too high a dose can cause kidney problems.

Some drugs alter the release rate of neurotransmitters. The more released by the sending neuron, the greater the chance that the message will get across strongly. Similarly, the more carrier pigeons released, the more are likely to deliver their message.

The availability of special sites, known as receptors, on the neuron receiving a message affects how well neurotransmitters deliver a message. Some drugs have the effect of blockading the receptor sites thus stopping or reducing the strength of the signals. Others, conversely, increase the availability of receptor sites, enhancing the communication. It is like the carrier pigeons being unable to get into their "home" (below) or having plenty of spaces to enter (right).

In the synaptic gap neurotransmitter levels are reduced by the action of enzymes that break them into non-active components, like the hunter killing carrier pigeons. With fewer neurotransmitters carrying the message, the communication becomes weaker. Some drugs have the opposite effect and inhibit the action of enzymes, thus raising neurotransmitter levels and boosting communication.

by altering levels of neurotransmitters in the brain, usually by boosting norepinephrine and/or serotonin.

In the 1950s, the only effective sedatives, or antianxiety drugs, were barbiturates, which are rarely used today because they are so addictive and potentially dangerous. Since 1960, the benzodiazepines such as Librium (chlordiazepoxide) and Valium (diazepam) have been available. These seem to have a more specific and subtle chemical action than barbiturates. But they are addictive and at one time were overprescribed. Their use now is limited to short periods only. Anxiety is also sometimes treated with antidepressants.

See also

STATES OF MIND
Talking cures 164/165
Feeling low 168/169
The mind adrift 170/171
Anxious states 172/173
Drugs of abuse 178/179

SURVEYING THE MIND
Discovering through damage 24/25
Probing the mind 26/27
Brain maps 30/37

BUILDING THE BRAIN
Crossing the gap 52/53
On or off 56/57
Unlocking the gate 58/59
Discovering transmitters 60/61

Feeling low

If sadness persists and deepens, people can feel trapped in the downward spiral that leads to depression.

Nearly all of us feel miserable and depressed sometimes – perhaps when a love affair ends, when a job is lost, or even just when the weather is wet. But for some, depression is not merely a passing mood; it develops into a way of life. They become overwhelmed by sadness, pessimism, and a sense of failure. They have difficulty concentrating and feel that any effort is futile. And they find it hard to get to sleep, still harder to wake up and often lose their appetite. It affects people from all walks of life, but manual workers and those from lower income groups are more likely to suffer than the middle classes, and more women suffer than men. About 15 percent of adults will have a severe episode of clinical depression at some time.

There are many theories as to the cause of depression, including genetic predisposition, the effects of childhood trauma, a chemical imbalance, and the idea that people become victims of "learned helplessness." They develop feelings of helplessness, perhaps because they have little control over their lives or because in the past they have been in situations where they could do little to relieve their distress.

A few people swing between deep depression and wild euphoria when they believe, perhaps, they can conquer the world. This condition is known as manic-depression, or bipolar disorder (in contrast to continuous unipolar, or clinical, depression), and seems to respond well to treatment with the drug lithium. Unipolar depression is much harder to treat, and psychiatrists deal with it using a combination of antidepressant drugs and psychotherapy.

Certain foods, including red wine and cheese, contain the chemical tyramine, which is normally broken down by the enzyme monoamine oxidase in the liver. If they are eaten when MOAIs are being taken for depression, tyramine accumulates, raising blood pressure beyond safe levels.

THE CHEMISTRY OF DEPRESSION

Varying levels of brain chemicals and patterns of brain activity seem to play a major part in depression. According to one theory, high levels of the brain chemical serotonin (one of the monoamine group) keep the brain in a state where it is at a low level of arousal – in effect, in a depressed state. Paradoxically, treatment involves raising serotonin levels even higher. This is thought to work by bombarding brain cells with so much of the chemical that they lose their sensitivity to it; cells thus become more active, arousal levels rise, and the depression lifts in about 10 days as the cells' sensitivity levels gradually decrease.

Two general classes of drugs raise serotonin levels: the first stops it from being reabsorbed and includes the tricyclics, a subgroup of the monoamine re-uptake inhibitors (MARIs), and the selective serotonin re-uptake inhibitors (SSRIs) such as fluoxetine (Prozac). The second class, the monoamine oxidase inhibitors (MOAIs), stops serotonin from being broken down, thus maintaining levels of the chemical.

HAPPY PILLS

No antidepressant drug has attracted more attention than fluoxetine, better known by its trade name Prozac. Since it was introduced in the late 1980s, use of the drug has soared in the United States. It seems to work with a variety of depressed patients, who typically show an improvement three weeks after being prescribed Prozac. The drug inhibits the re-uptake of serotonin, boosting levels of this chemical in the brain and having the net effect of moving it into a higher state of arousal.

Many depressives swear that Prozac has been their salvation, and even those who are not clinically depressed use it to help liven up their social life. For some it has become a way of life. But the drug has unpleasant side effects – such as nausea and agitation – and it is also addictive. Critics argue that Prozac pushes people into a state of perpetual mindless sensuality, with no thought for others.

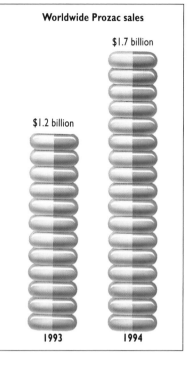

Worldwide Prozac sales

$1.7 billion

$1.2 billion

1993 · 1994

Depression is a natural reaction at times, for example after the death of a loved one. In such instances the cause of the depression is obvious – but it is not so clear why some people recover rapidly and others do not. Whatever the case, the grieving process takes time and should not be hurried; neither should the fact of a loved one's death be hidden away or ignored. Bereavement is a common experience: every year more than 8 million people in the U.S. experience the death of a member of their immediate family.

Where there is an identifiable cause, such as bereavement or physical illness, psychiatrists talk of exogenous depression. Where there is no such obvious cause and the depression appears to be coming from within, it is known as endogenous. Even so, this type of depression may have an external cause – perhaps buried in the person's past – that is simply hard to identify.

Energetic sports such as aerobics may be a simple way of helping to fight clinical depression. Most people experience a feeling of wellbeing as a result of strenuous physical activity because it causes the brain to release endorphins, natural painkillers that help to alleviate discomfort and stress. And some psychiatrists think that if patients follow an increasingly demanding daily schedule of activities – especially social activities – they are likely to become more confident, which will assist in lifting their depression.

See also

STATES OF MIND
▶ Emotional states 154/155

▶ Abnormal states 162/163

▶ Talking cures 164/165

▶ Physical cures 166/167

▶ Anxious states 172/173

▶ Fears and fixations 174/175

SURVEYING THE MIND
▶ Map of chemicals 36/37

BUILDING THE BRAIN
▶ Unlocking the gate 58/59

▶ Discovering transmitters 60/61

INPUTS AND OUTPUTS
▶ Rhythms of the mind 120/121

FAR HORIZONS
▶ Learning to be human 144/145

The mind adrift

A number of factors – genetic, physical and environmental – may contribute to schizophrenia.

Sometimes, the inner world of the schizophrenic can be confusing and even terrifying. A cacophony of sensations bombards the mind; angry inner voices shout orders; gibberish trips off the tongue. Yet the symptoms are not always so extreme. Indeed, they vary so much, and the line between schizophrenia and normality can be so blurred, that it is often impossible to diagnose with complete certainty.

Schizophrenia is one of the most serious psychiatric disorders. Usually symptoms are divided into two types: positive (Type I) and negative (Type II). Positive symptoms include hearing voices, suffering from delusions and hallucinations, and behaving strangely; they are more dramatic, but tend to be shortlived. Negative symptoms include apathy, depression and confused speech; they last longer and occur at more often, some think, as the illness tightens its hold. Schizophrenia typically strikes in early adulthood and total recovery after a major attack is rare. Positive symptoms can be eased by tranquillizers and psychotherapy, but the victim's ability to carry on normally between attacks tends to decline.

Brain cross section

Normal ventricles   Enlarged ventricles

Striatum

Frontal cortex

Ventral tegmental area

Amygdala   Substantia nigra

Dopamine pathways in the brain

The brain chemical dopamine is implicated in schizophrenia. Major dopamine pathways go from the substantia nigra to parts involved in movement and from the ventral tegmental area to areas that link sensory perceptions with memories and emotions.

Schizophrenics are often acutely aware of the bewildering array of information pouring into their brains all the time from their senses. Most people manage to filter out some of these sensations – just as you mentally block out background sounds in a noisy café and focus on your friends' voices. But schizophrenics are unable to do this. So they find their minds constantly bombarded by intense sights and sounds.

Physical abnormalities in the brain can be linked to schizophrenia, but it is not certain whether these signs of damage cause the disorder or are caused by it. MRI scans reveal larger ventricles in the brain of a schizophrenic twin than in his or her healthy identical sibling (left). PET scans show that the prefrontal cortex – crucial to forming memories – is much less active in sufferers. So, too, is the hippocampus, which is also important in memory. Post mortems have shown that the prefrontal cortex of schizophrenics is often atrophied, and this shrinking of brain tissue seems also to enlarge some of the cavities or ventricles.

See also

STATES OF MIND
▶ Abnormal states 162/163

▶ Talking cures 164/165

▶ Physical cures 166/167

SURVEYING THE MIND
▶ The working mind 16/17

▶ Inside the mind 20/21

▶ The view from outside 28/29

▶ Brain maps 30/37

BUILDING THE BRAIN
▶ Discovering transmitters 60/61

INPUTS AND OUTPUTS
▶ The big picture 114/115

FAR HORIZONS
▶ The infinite store? 130/131

WORD SALAD

A woman was delighted to receive a letter from her son abroad, but distraught when she read it: "Dear mother ... I am writing on paper. The pen I am using is from a factory called Perry and Co. The factory is in England. The city of London is in England. I know this from my school days. Then I always liked geography. My last teacher in that subject was Professor August A. He was a man with black eyes. There are also blue and grey eyes and other sorts too. I have heard it said that snakes have green eyes. All people have eyes. There are some, too, who are blind."

He was later diagnosed as being schizophrenic. Sufferers often have difficulty using language to communicate. They cannot make small talk and tend to answer questions with a single word. Although they do not lose their linguistic ability – their word play can appear to border on genius – they lose the power of abstract thought and seem to flit from one phrase to the next by random association without any shaping strand. One word may follow another simply because it rhymes or conjures up particular images. This may also be because working memory, used to help us speak sentences, is affected. A schizophrenic will not retain the memory of the beginning of a sentence and so finishes it with nonsense.

Genetic factors may play a role in causing schizophrenia. Studies have shown that the disorder is more likely to develop in the relatives of schizophrenics than in those of non-schizophrenics, and the closer the relationship, the greater the chance of having it. For example, if one identical twin suffers from schizophrenia, the other twin (whose genes are identical) has a 40–60 percent chance of developing the illness. With non-identical twins, the likelihood drops to 17 percent.

The hunt is on for the guilty gene. By analysing the DNA of at least 200 pairs of afflicted siblings, scientists hope to pinpoint the position of genes that could be linked to the disorder. Some researchers also think that a genetic defect could disrupt the migration of neurons in the developing brain, leading to abnormal connections between the cortex and the limbic system, thought to play a role in emotion and motivation. The illness occurs to the same degree worldwide and it still exists, suggesting that the gene that carries it must have other evolutionarily important factors.

Environment can be the trigger for schizophrenia. Emotional strife in the family is one such factor. In "double-bind" situations, a parent gives contradictory messages. A child may be told "don't be so obedient", for instance. Where there is marital discord, a child may be encouraged by mother or father to side with them, thus fostering guilt feelings about split loyalties.

Usually children are taught how to respond to social cues – to a frown or smile – and to moderate behaviour accordingly. If this does not happen, they may concentrate on irrelevant cues such as the hum of a computer.

Scientists believe schizophrenia is linked either to raised levels of the neurotransmitter dopamine in the brain or to increased sensitivity of dopamine receptors, both of which induce a high state of arousal.

Symptoms can be treated by drugs like phenothiazine which block dopamine receptors in brain cells and so reduce the chemical's effect. Conversely, drugs which boost dopamine levels, such as L-dopa (given to Parkinson's sufferers), create schizophrenia-like symptoms in healthy people. There are five different dopamine receptors, D1 to D5. Schizophrenics have more D2 receptors in the basal ganglia part of the midbrain, but disruption is not limited to this area. The problem may be caused by changes in certain fats in neuron membranes, which alter the way dopamine works.

171

Anxious states

It is natural to feel worried sometimes; but for a number of people, anxiety can be both disabling and distressing.

From time to time, all of us get worried or anxious – before a major exam, or when a loved one is late coming home, for instance. But for some people, anxiety can become so extreme, or can be provoked by so little, that they are said to be suffering from anxiety disorder. They become tense, nervous, and distressed, and sometimes panic. They may spend a lot time worrying about why they are so anxious, and then become frustrated because they cannot see why. And they may be beset by a whole range of somatic, or physical, symptoms. Some break out in a sweat; others tense their muscles, possibly causing headaches; still others hyperventilate, which makes them dizzy, or develop high pulse rate or blood pressure – all signs of a body aroused by fear for "fight or flight." Indeed, each individual's anxiety is expressed with a slightly different range of physical symptoms, from stomach cramps to susceptibility to colds – or simply pacing around the room or tapping fingers.

There are as many kinds of anxiety disorder as there are anxious people, but five main categories can be identified: phobias, obsessive-compulsive disorders, panic attacks, generalized anxiety, and stress disorders. The last of these includes post-traumatic stress syndrome, which occurs when, after a terrible event such as a fire or car accident, victims suffer from recurring painful memories, nightmares, and flashbacks, often so vivid that they feel they are reliving the event.

For some people, a particular situation seems to induce greater anxiety levels. The reason for this is not known, but a genetic predisposition and how they were treated in childhood are possible theories. Panic attacks can come from out of the blue, with no apparent cause. According to one theory, they are caused because a person is highly sensitive to a few physical sensations and thinks they foretell the failure of some vital bodily function. Another attributes them to the emergence of long-repressed painful memories or difficult emotions.

Anxiety disorders affect 15 percent of people at some time. Since they are more common in technologically advanced societies, they are thought to be caused by the stress of modern living – noisy cities, fast pace, the breakdown of communities and relationships, urban isolation, and competitive lifestyles.

Singer Joan Baez suffered from severe stage fright *in the past. The symptoms can include breaking out in a cold sweat or getting "butterflies" in the stomach before going out in front of an audience. These stage fright symptoms are part of the body's automatic responses to anxiety. Joan Baez dealt with this problem, and others, using psychotherapy, yoga, and visualization techniques, among other things.*

Stress affects us physically *because we react to it with the "fight or flight" response. In preparation for either staying and fighting or fleeing from danger, heart and breathing rates speed up to get more oxygen to the brain; the digestive system "shuts down" temporarily to avoid wasting energy; and skin temperature drops, because of sweating and the constriction of small blood vessels in the skin, turning it paler and redirecting blood to the muscles.*

It is thought that these physical responses may be in part responsible for stress-related heart disease. One theory is that since stress narrows blood vessels and boosts heart rate, it puts the coronary arteries under strain. Another is that hormones activated by stress make blood pressure waver, weakening blood vessels. A third is that stress adds lipids (fats) to the bloodstream, clogging the arteries. To reduce the risk of heart disease, people are advised to modify their reaction to stress.

Anxiety disorder can be triggered by stress, which may make us more likely to suffer from all kinds of physical illness. Stress is essentially caused by a situation in which we feel challenged or threatened in some way – one in which we must either adapt or cope. Things that cause stress – called stressors by psychiatrists – can range from missing a train to buying a house.

Various studies have rated the stress level of different life events. Such ratings are inevitably rather arbitrary – events which are stressful for some are not for others – but they give an idea of how things might affect us. Since the late 1960s, psychiatrists in the U.S. have used the SRRS (social readjustment rating scale), which gives scores to 43 stressors that affect our lives. At the top of the list, with a tally of 100, comes the death of a husband or wife. Next comes divorce (73), a broken relationship (65), a prison sentence (63), the death of a close family member (63), and personal injury or illness (53). Marriage comes next on the list with a score of 50. Down near the bottom of the scale are vacations (13) and Christmas (12), which are both in theory pleasant, but can also cause some stress.

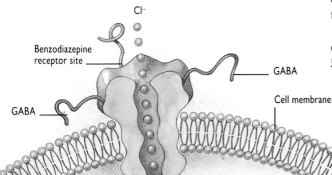

Learning to relax can be an effective way of combating stress. Even if you cannot avoid stressful situations, you may be able to modify your reaction to them. Some people suggest relaxing by exercise – sports, aerobics, dancing, and so on – or transcendental meditation, which has been shown to reduce stress and improve physical health. Something as simple as walking the dog can also help. And letting your emotions out or even just laughing can have a restorative effect. Others suggest a cognitive approach – that is, modifying your way of thinking. If you get stressed when someone cuts in front of you in the car, for instance, remind yourself that it really makes very little difference.

See also

STATES OF MIND
▶ Abnormal states 162/163

▶ Talking cures 164/165

▶ Physical cures 166/167

▶ Feeling low 168/169

▶ The mind adrift 170/171

▶ Fears and fixations 174/175

BUILDING THE BRAIN
▶ On or off 56/57

▶ Unlocking the gate 58/59

▶ Discovering transmitters 60/61

▶ Remaking the mind 62/63

INPUTS AND OUTPUTS
▶ Keeping control 116/117

▶ Dealing with drives 118/119

Buspirone

5HT₁ₐ autoreceptor

Cell body

Axon

5HT–producing cell

5HT

THE CHEMISTRY OF ANXIETY TREATMENTS

Clearly, stress and individual personality play a major part in anxiety disorder, but some physical explanations have also been suggested. The overall levels of certain brain chemicals – such as GABA (gamma-aminobutyric acid) and serotonin (5HT) – are thought to be important. GABA, for instance, has an inhibitory effect; it reduces the sensitivity of neurons, so it takes more to excite them. It may be deficient in anxious people, allowing the brain to become overaroused and causing anxiety. So raising GABA levels or enhancing its effect might reduce anxiety. The benzodiazepine (BDZ) tranquillizers, such as Librium (chlordiazepoxide) and Valium (diazepam), work partly by increasing GABA activity. At the cell-membrane level, BDZ binds to a receptor site on a GABA receptor and enhances the effect of GABA, causing the GABA chloride channel to open more often. The influx of chloride ions hyperpolarizes the cell, making it less likely to fire.

Another antianxiety drug is buspirone, which has the net effect of reducing the output of 5HT in the brain. Lowering 5HT levels is thought to reduce anxiety. Buspirone binds as an agonist to an autoreceptor, known as the 5HT₁ₐ autoreceptor, on a 5HT-producing cell. An autoreceptor regulates the output of a neuron, acting a bit like the thermostat on a electric iron. If the cell produces too much neurotransmitter, in this case serotonin, the amount binding to its autoreceptors rises, telling the cell to cut its output.

Cl⁻

Benzodiazepine receptor site

GABA

GABA

Cell membrane

GABA

Fears and fixations

There are many unfortunate twists and turns in the paths of human thought – some are truly terrifying.

Imagine stepping out onto the balcony of your vacation hotel. Suddenly, it shakes violently, the hand rail disintegrates, the floor dissolves, a bottomless chasm gapes open beneath your feet, and you start to fall. Naturally you are terrified. This is what going out on a balcony is like all the time for someone who suffers from acrophobia, an abnormal fear of heights. Acrophobia is just one of many phobias – abnormal fears – that may afflict some people. Phobias are not unusual – according to some research, about 8 percent of women and 3.5 percent of men suffer from a phobia at some time.

Phobias are essentially excessive, unreasonable fears, in which someone is terrified of something – a situation, activity, or object – that holds little or no danger. A phobia becomes a problem when it affects a person's ability to live a normal life, or when it makes someone extremely unhappy or anxious. When this is the case, the person is said to suffer from a phobic disorder. Phobias break down into three broad categories. There are simple phobias: fear of such things as spiders, water, darkness, birds, and so on. There are social phobias: fear of being criticized by others, which makes

Unwanted thoughts and irresistible urges, which can be neither understood nor controlled, are thought to beset about 1 person in 40. Obsession is when a person's mind becomes dominated by an image or impulse which the sufferer cannot get rid of. The thought takes over the person's life and guides his or her actions – and seems impossible to fight against.

Compulsion is when someone is overcome by an irresistible urge to perform a meaningless act repeatedly. Some people become so obsessed by cleanliness, for instance, that they compulsively wash their hands hundreds of times a day. If these thoughts or actions – either alone or together – inhibit their everyday life or make them unduly anxious, then they are said to have obsessive-compulsive disorder.

Obsessive-compulsive disorder may have something in common with "stereotyped" behavior in animals. This occurs when cagebound animals repeat the same movement again and again in a highly agitated manner with no apparent purpose.

A different use of the word obsession describes the sexual obsession of fetishism. The key feature is that sexual urges and arousing fantasies are fixated on an inanimate object or unexpected body part. A fetishist might think of or handle the object while masturbating or, if the object is an item of clothing, would not be able to have intercourse unless a sexual partner were wearing it. The range of fetishistic objects is vast; in fact, almost anything can be a fetish – from wrists to rubber.

EATEN UP BY FEAR OF FAT

Thousands of people – usually teenage girls – suffer from one of two major eating disorders: anorexia nervosa and bulimia nervosa, both of which are serious (5–20 percent of anorexics die from the damage done to their bodies). Recently, the number of males with anorexia has increased; some put the ratio of male to female sufferers at 1 to 10. Anorexics refuse to eat enough – often to the point of starvation. No one knows what causes the disease, and it is difficult to treat. Most sufferers see themselves as fat – even when emaciated – and stubbornly refuse to eat. Bulimics are similarly concerned about their weight, but they alternate binges, when they stuff themselves with food, and purges, when they get rid of it with laxatives or by making themselves sick. It is also hard to treat, since bulimics often isolate themselves to hide their behavior.

A common phobia is arachnophobia, an irrational fear of spiders. People can develop irrational fears of all kinds of things, and there are conflicting theories as to how phobias come about. The behaviorist John B. Watson carried out a famous, if by today's standards unethical, experiment early in the 20th century to show how phobias might be learned by association of genuinely frightening stimuli with "harmless" ones. He gave a year-old baby named Albert a white rat to play with and then frightened the infant by banging on a metal bar. After repeating this several times, Albert became frightened of the rat. He also became afraid of white dogs and rabbits. But there is some debate as to whether his reactions were as strong as a phobia, and the case for phobias being caused by this type of association, or conditioning, is not proven. Other theories state that phobias arise more often when society is unstable; that they are manifestations of defense mechanisms to control underlying anxiety; or that a predisposition to certain phobias is built into the human mind by evolution. But none of the theories seems able to explain phobias fully.

people petrified of talking to the opposite sex, eating in public, meeting new people, and so on; these are rarer than simple phobias. And there is agoraphobia, the most common of all, especially for young women. Agoraphobia, literally, fear of the market place, is usually defined as fear of open spaces, and some agoraphobics are indeed terrified of being alone in wide open spaces. But for most it is a fear of being in public places, especially if they do not have a friend or family member with them. Many are afraid that they might not be able to escape from such a place if they lost control of some bodily function or fainted. When severe it can confine people to their homes.

Usually, phobics are completely aware of how irrational their fear is and are often upset by their inability to overcome it. Some are successfully treated by hypnosis, but the medical profession usually tries behavior therapy. In systematic desensitization, a phobic's ability to cope with the fear is built up by gradual exposure to more and more of what frightens him or her. Someone who was terrified of spiders might initially be asked to think about a spider, then to look at a picture of one, and so on, building up to being able to handle a spider. The opposite approach is taken in "flooding" – exposing the phobic to their fears. Thus taking an acrophobic to the top of a high building or a nyctophobic into a darkened room forces them to realize that nothing bad happens to them. "Implosion therapy" is midway between the two, encouraging phobics to live through their worst fears – but only in their imagination.

NAME THAT PHOBIA

Animals	Zoophobia
Being afraid	Phobophobia
Being alone	Autophobia
Being buried alive	Taphophobia
Being dirty	Automysophobia
Being stared at	Scopophobia
Birds	Ornithophobia
Blood	Haematophobia
Body odors	Osphresiophobia
Books	Bibliophobia
Cancer	Carcinomatophobia
Cats	Ailurophobia
Choking	Pnigophobia
Crowds	Ochlophobia
Darkness	Nyctophobia
Death	Thanatophobia
Dirt	Mysophobia
Disease	Nosophobia
Dogs	Cynophobia
Enclosed spaces	Claustrophobia
Failure	Kakorraphiaphobia
Feces	Coprophobia
Fire	Pyrophobia
Flying	Aerophobia
Food	Sitophobia
Foreigners	Xenophobia
Germs	Spermophobia
Heights	Acrophobia
Human beings	Anthropophobia
Insanity	Lyssophobia
Insects	Entomophobia
Light	Photophobia
Machinery	Mechanophobia
Men	Androphobia
Motion	Kinesophobia
Nakedness	Gymnophobia
New	Neophobia
Noise or loud talking	Phonophobia
Open spaces	Agoraphobia
Pain	Algophobia
Physical love	Erotophobia
Pleasure	Hedonophobia
Poverty	Peniaphobia
Punishment	Poinephobia
Reptiles	Batrachophobia
Robbers	Harpaxophobia
School	Scholionophobia
Sea	Thallassophobia
Sharp objects	Belonophobia
Sleep	Hypnophobia
Spiders	Arachnophobia
String	Linonophobia
Thunder	Keraunophobia
Travel	Hodophobia
Vomiting	Emetophobia
Water	Hydrophobia
Women	Gynophobia
Words	Logophobia
Work	Ergasiophobia
Writing	Graphophobia

See also

STATES OF MIND
▶ Emotional states 154/155

▶ Abnormal states 162/163

▶ Talking cures 164/165

▶ Physical cures 166/167

▶ Feeling low 168/169

▶ Anxious states 172/173

▶ Motivation 180/181

SURVEYING THE MIND
▶ The working mind 16/17

FAR HORIZONS
Conditioning the mind 128/129

Learning to be human 144/145

The failing mind

Most people have a fully functioning mind all their life; a few, however, slowly lose their higher faculties.

Nowhere is the connection between brain cells and all the processes that we think of as making us uniquely human made more clear than in cases of dementia. In such cases, over years, sometimes decades, the brain cells which play a part in understanding, memory, and language slowly die off until there is little of the mind left that is recognizably human. When the degeneration begins, sufferers become detached from friends and relatives. Later, over half of them become paranoid and believe they are being persecuted. Day-to-day activities, such as using the telephone, writing checks, and driving, become increasingly difficult. Later still, patients need help in feeding, dressing, washing, and going to the bathroom.

Dementia is largely a disease of old age, affecting only one percent of 60 to 64-year-olds, but 30–40 percent of those over 85. It can be caused by blood clots in the brain as well as by Parkinson's disease. The hallmark of Parkinson's is shaking hands and a shuffling walk, but 40 percent of sufferers can also experience severe intellectual decline. It is caused when cells in the substantia nigra of the midbrain, which are involved in controlling movement, start to die off.

But the leading cause of dementia is Alzheimer's disease, accounting for 50–60 percent of cases. In the brains of sufferers, whole areas are filled with deposits, or plaques, and tangled whorls of protein that damage the surrounding neurons. There is no known cure, but in some countries the drug tacrine, which blocks the enzyme that destroys the neurotransmitter acetylcholine, has boosted intellectual performance in the short term. The cause is still unknown, but people are more likely to succumb if members of their family have had it, if they have suffered a head injury, or if they have Down's syndrome.

Genes that seem to be associated with Alzheimer's have been discovered. One such is FAD3 – familial Alzheimer's disease gene no. 3 – linked to the uncommon early-onset form. Carriers of an abnormal FAD3 gene start sliding into dementia in their 40s. An increased chance of developing the more common late-onset form of Alzheimer's is linked with a mutated gene for the protein ApoE, normally involved in transporting fats in the blood. Just how these genes have their effect is still unclear.

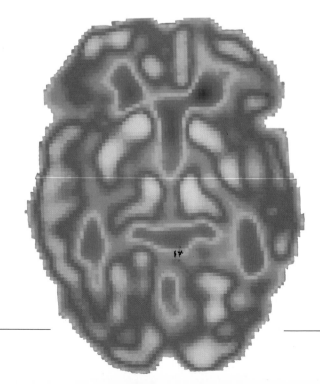

Alzheimer's disease is no respecter of rank or privilege. It is now official that former president of the United States Ronald Reagan suffers from this debilitating illness. The man called "the great communicator" *made one of his last public appearances, with his wife Nancy, at a gala in Washington to celebrate his 83rd birthday in February 1994.*

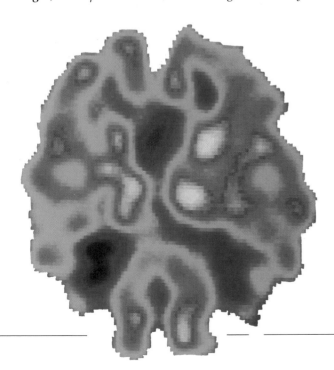

PET scans show brain activity levels from low (blue) to high (yellow). The normal brain (**below left**) shows symmetrical high activity in the left and right cerebral hemispheres. The scan of an Alzheimer's sufferer (**below right**) shows patchiness and reduction in general activity.

THE LOSS OF A MIND

The relatives of an elderly woman living alone became worried when her neighbors complained, on several occasions, about the smell of burning from the woman's apartment. She often broiled her food but became prone to forget that she had put it under the heat, where it caught fire. The problem grew steadily worse until she was no longer able to look after herself and had to move into a nursing home. There she stopped talking, did not recognize her relatives, and eventually died after some years. When the woman began to burn her food, she was in the early stages of Alzheimer's disease.

As people age, plaques and tangles begin to build up in the brain, especially in the hippocampus, an area that is involved in laying down new memories. These lumps or plaques of protein in the brain and tangles of nerve filaments gradually damage the surrounding neurons, leading to forgetfulness, but they do not, to any real extent, affect the power of the mind.

In Alzheimer's disease, the plaques and tangles are far more numerous than usual and are found all over the brain, including in the cerebral cortex – the realm of higher functions. At the heart of the plaques is a short chain of protein, known as beta amyloid, which has been chemically severed off a much larger protein – beta amyloid precursor protein. Normally, the beta amyloid is cut in two, but the trouble starts when, for some reason not yet understood, the enzymes isolate intact beta amyloid, which is then liberated and starts to cause damage.

The progress of dementia is marked by a thinning of the cortex. While the medial temporal lobe (MTL) of a normal healthy aging person thins gradually over time, the MTL of someone with Alzheimer's or dementia of an Alzheimer's type shrinks rapidly on average from the time of diagnosis. Death usually comes when the brain in this region is ⅛ inch (3 mm) thick.

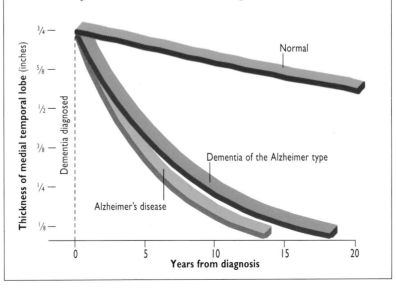

See also

STATES OF MIND
▶ The unique self 156/157

▶ Abnormal states 162/163

▶ Individual minds 184/185

SURVEYING THE MIND
▶ Discovering through damage 24/25

▶ Brain maps 30/37

BUILDING THE BRAIN
▶ Support and protection 42/43

▶ The aging brain 76/77

FAR HORIZONS
▶ The infinite store? 130/131

Drugs of abuse

Most people use drugs daily – to wake them up, to relax, to get a buzz – even if they don't classify them as such.

It seems such a modern problem, yet people have been taking drugs to alter their mood since the dawn of civilization. Today, drug taking is the norm: many people drink tea, coffee, and alcohol; many smoke cigarettes; many young people have tried illegal drugs such as marijuana at least once in their lives; and a huge number take tranquillizers or antidepressants.

Yet however commonplace they may seem, a lot of drugs can be harmful. Some, such as tobacco, cause huge physical damage if used heavily over long periods. Others can do psychological damage. Frequent use of amphetamines can bring on paranoia and psychosis similar to schizophrenia. Heavy drinkers suffer both physical (cyrrhosis of the liver) and psychological (Korsakoff's syndrome, the inability to store memories) damage: But many drugs – including alcohol – are only really harmful if taken in excess. However, most people could not argue with the idea that if either an individual or society is harmed by a person's taking of a drug, then the drug is being abused. When abuse has been perceived as a threat at various times and in different cultures, certain drugs have been declared illegal.

*A **house of cards** is an exceptionally delicate structure and must be perfectly balanced if it is to remain standing. The slightest puff of wind will have the cards cascading to the ground. The brain is an infinitely more complex structure whose equilibrium is easily disturbed, and drugs that interfere with the chemicals that send signals within it can have catastrophic effects on that balance.*

People become addicted to drugs *such as cocaine (snorted as "coke" and smoked as "crack"), heroin, and alcohol when the body becomes dependent – the drugtakers cannot function normally either physically or psychologically without the drug. If they stop taking the drug, they suffer withdrawal symptoms – often so hard to bear that they feel compelled to seek immediate relief in the drug again. This is why addicts find it so hard to wean themselves off the drugs.*

Slightly less serious than dependence is tolerance, when the body adapts to the continual presence of the drug. As a result, the drugtaker has to take ever larger quantities to achieve the same effect. This happens quite frequently with drugs such as alcohol and Ecstasy and may lead to addiction.

When a person becomes addicted, or even just tolerant, the problems of obtaining and paying for the drug are added to the physical and psychological dangers of excess. This results not only in potential harm to the individual, but also in danger to society.

See also

STATES OF MIND
▶ Abnormal states 162/163

▶ Physical cures 166/167

▶ Feeling low 168/169

▶ The mind adrift 170/171

▶ Fears and fixations 174/175

▶ The moral mind 182/183

BUILDING THE BRAIN
▶ Discovering transmitters 60/61

INPUTS AND OUTPUTS
▶ Feeling pain 104/105

▶ Keeping control 116/117

PSYCHOACTIVE DRUGS

Drugs which alter mood, behavior, and consciousness – known as psychoactive drugs – come in many forms, but all contain substances such as the caffeine in tea or the alcohol in beer which alter the brain's chemistry in some way. Some boost levels of a particular chemical, for instance; others make a certain brain chemical act more or less effectively or even mimic it. By altering the brain's chemistry, these substances change our state of mind. Such drugs can be grouped into four categories – narcotics, depressants (downers), stimulants (uppers), and hallucinogens (psychedelic drugs) – and although numerous types already exist, chemists are searching for new ones all the time.

Narcotics include the original narcotic opium, made from the poppy, as well as other opiates – natural narcotics such as heroin, morphine, and codeine – and opioids, or synthetic narcotics, such as meperidine and methadone. The word narcotic comes from the Greek for "numbness," and these drugs are the most powerful of all painkillers. Drug abusers take them for the feeling of relaxed euphoria, the "high," they give. They work by mimicking endogenous morphines or endorphins, the body's natural painkilling neurotransmitters, heightening the pain relief and euphoria of a natural release of endorphins. The more narcotics a person takes, the fewer endorphins the body makes, increasing tolerance and dependence. Eventually, natural endorphin production falls so low that any reduction in drug dosage brings horrible withdrawal symptoms – chills, sweating, stomach cramps, headaches, and vomiting. Worse still, the body's greater tolerance for the drug spurs the abuser to take ever higher doses to achieve a high, so increasing the risk of a fatal overdose.

Depressants, including alcohol, sedative-hypnotics like barbiturates, and tranquillizers, do not make you depressed; they depress or slow down the activity of the central nervous system, which is why some of them are known as downers. Small doses raise your mood, relax you, and free you of inhibitions. High doses can make you moody, anxious, and irritable. They also slow your reflexes, slur your speech, and upset your balance and judgment. Alcohol, for instance, seems to work by reducing the effectiveness of dopamine and other neurotransmitters. In small doses, it excites you because it slows down only the inhibitory nerve pathways – those that keep you in check. But high doses make you drunk because they also slow down the nerve pathways that excite and arouse you.

Stimulants include substances such as caffeine, nicotine (in tobacco), amphetamine (speed), cocaine, and MDMA (3,4-methylenedioxymethamphetamine, or Ecstasy). In small doses, they wake you up and give you a high by exciting the central nervous system. In high doses, they can make you anxious and irritable and even psychotic. All of them can be addictive.

Caffeine is the drug in many national drinks – coffee in the U.S., tea in Britain, guarana in Brazil – and in chocolate and cola drinks. It perks you up by stimulating the heart and suppressing the effects of adenosine, one of the brain's inhibitory chemicals. Speed arouses you even more, inducing a high by boosting levels of the brain chemicals norepinephrine and dopamine, and keeping them in action longer by preventing their re-uptake from the synapse by nerve cells. High doses also affect levels of serotonin. Cocaine works in much the same way, but affects norepinephrine and serotonin more, and gives a brief high, rather than long arousal. Ecstasy gives a feeling of euphoria and sociability, and seems to work directly on the brain's serotonin pathways, slowing the re-uptake of the chemical from synapses.

Nicotine seems to stimulate respiration and heart rate and depress appetite by activating nicotine-sensitive nerve receptors and mimicking transmission of the brain chemical acetylcholine. It is highly addictive, and the dangers it poses to health are well documented.

Hallucinogens alter the way a person perceives things, either mildly like marijuana or dramatically like "acid" (LSD; lysergic acid diethylamide), mescaline, psilocybin (magic mushrooms), and PCP (phencyclidine), which all induce hallucinations. Most (but not marijuana, PCP, or mescaline) affect the serotonin system that reaches up from the brain stem to the rest of the brain, which is why they have such a profound effect on consciousness. One theory is that serotonin usually stops us from dreaming when we are awake, and inhibiting the serotonin fountain allows us to do just that. But these drugs affect people differently, giving some a wonderful dream and others a nightmarish bad trip.

Marijuana seems to affect the brain chemical GABA (gamma-aminobutyric acid). It may induce mild relaxation, but it can also make time seem to pass slowly or even cause hallucinations.

Motivation

Why do you get up in the morning? What makes you go to work or school? What makes you do anything?

Questions of motivation – why people do what they do – are among the most difficult to answer. Early this century, psychologists such as William James tried to explain them in terms of "instinct" – automatic responses inherited from our parents, including "physical" instincts such as sucking and "mental" ones such as jealousy. Everything we do would be motivated by a different instinct. But as the list of instincts rose over 10,000, theorists realized this approach was not quite right.

Then psychologists came up with the idea of "drives" – compelling desires to satisfy the physical need for such basics as food, water, sleep, warmth, and sex. Ultimately, some argued, everything we do could be understood in terms of an attempt to satisfy, or reduce the clamor of, these drives. One problem with this theory is that humans and other primates are naturally curious, exploring for the joy of it, and this is hard to explain in terms of physiological needs. Now psychologists simply try to explore the links between brain physiology and motivation, and to develop theories of our needs and ways of thinking.

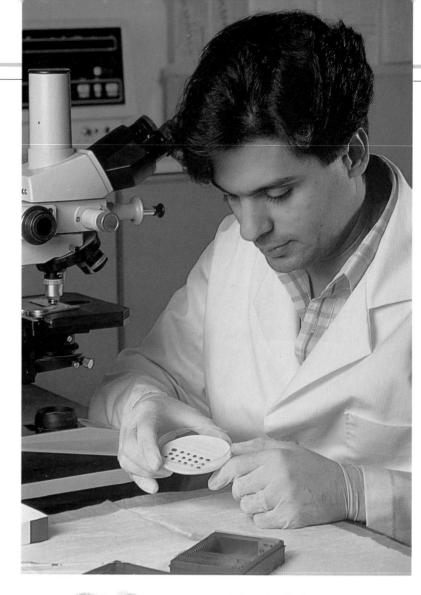

People talk about the sex drive as if it were a basic physiological need like the need for food. But while people have to eat to avoid starvation, a lack of sex will not kill anyone. Yet people still seek out sexual pleasure even though it is not essential to their immediate survival. Sexual desire is linked with many motivations, including the needs for love, power, and physical closeness, to name but a few. Although they are different types of drive, both hunger and sexual desire are influenced by the hypothalamus in the brain. It stimulates the pituitary gland to release the sex hormones. When levels of these hormones fall, so does sexual desire.

The stylish lemon squeezer (left) represents a common human trait, the desire to create surroundings full of aesthetically pleasing things, from everyday implements to works of art.

Many people are highly motivated to acquire wealth and the possessions that it can buy. And there are few who could honestly say that they do not gain pleasure from material objects. What these are depends largely on the values of the culture in which a person lives. In some parts of the world, a powerful car is the object of desire, while in others it might be expensive jewelry. Objects can be perceived to bestow status on their owners and thus raise their self-esteem.

The scientist's constant quest for knowledge is one of the many examples of human curiosity – a human activity which cannot be explained in terms of physiological needs. Some psychologists argue that we are curious because we are trying to control our environment. But other factors may motivate people to achieve in this field. Peer-group admiration and the knowledge that your work might benefit society can both be immensely gratifying.

PRIMARY NEEDS

A mother's love for her offspring is so common among all species of animals that it is easy to understand in terms of instinct. Our needs for food and drink, too, are easily understood in terms of basic drives. But what about everything else we do? Many psychologists think we are motivated as much by social needs as by physiological drives and instinct. They argue that we are prompted by, for instance, a need for affiliation, which makes us seek out friends and join groups, or a need to achieve, which makes us look for challenges. The need to achieve is common to all of us, some believe, but its form varies from culture to culture – and from person to person.

Psychologist Abraham Maslow (1908–70) suggested that we are motivated by a need hierarchy: only when we have satisfied the lower levels of needs do we try to satisfy the higher ones. At the bottom are physiological needs such as those for food and drink. Above that is the need for safety and security. Then comes the need to belong and be loved. Next is the need for self-esteem. Above that are so-called cognitive needs, which include exploration, curiosity, and the search for meaning and knowledge, followed by aesthetic needs for beauty, order, and symmetry. And at the top comes the need for "self-actualization" – the need to fulfill our true potential.

The sow with her piglets is highly protective of her offspring. She behaves in a way that promotes their wellbeing and enhances their chances of growing up and eventually reproducing. Although what she does might seem selfless – she looks after her offspring often to her own detriment – it might be seen as the satisfaction of a longer-term biological need. Evolution is driven by the survival of the fittest, especially those who are most successful at reproduction. Thus, successful nurturing behavior becomes part of the genetic inheritance as instinctive behavior.

Scientists used to think that we felt hungry when our stomach contracted. Now they know the brain, and in particular the hypothalamus, plays a major role, too. The lateral hypothalamus (LH) seems to be a switch that makes us want to eat; the ventromedial hypothalamus (VMH) tells us to stop eating. The LH and VMH may respond to changing levels of nutrients in the blood – the LH going into action when levels drop, and the VMH coming into play when they rise again. One theory suggests that the hypothalamus tries to keep each of us at a particular set weight, established by the number of fat cells in the body when we are born. Extreme obesity may be caused by damage to the hypothalamus.

The regulation of body temperature is a physiological essential: if we get too hot or too cold, we die. Automatic temperature regulation is controlled by the hypothalamus: we sweat to cool down and shiver to warm up. We also control body temperature voluntarily, finding shade when the sun is hot and dressing warmly when the air is cold. But some people deliberately choose hot sunny vacation destinations. This is not because they are cold – they have other associations with the sun, such as the pleasure of freedom from everyday cares.

See also

STATES OF MIND
▶ Emotional states
154/155

▶ The moral mind
182/183

SURVEYING THE MIND
▶ The working mind
16/17

▶ Inside the mind
20/21

BUILDING THE BRAIN
▶ Food for thought
66/67

INPUTS AND OUTPUTS
▶ Keeping control
116/117

▶ Dealing with drives
118/119

FAR HORIZONS
▶ Having ideas
140/141

▶ First thoughts
142/143

▶ Learning to be human
144/145

▶ Thinking together
150/151

The moral mind

You know when you are doing wrong, or at least you should if you have developed a set of morals.

News headlines trumpeting "Crime wave brings terror to city streets!" make us feel crime is too much the norm. Yet most of us behave well nearly all the time, and few would even think of stealing an ice cream from the hands of a child. Even though our appetites may tell us to grab the goodies, our minds stop us. Most of us have a strong moral sense and are essentially altruistic. Why?

The problem for scientists is that our altruism is apparently at odds with Darwin's theory of evolution, which shows how species evolve as the fittest (best-adapted) individuals survive to pass on their genes. Some have tried to apply this notion to human nature and society, with terrifying results: the idea that some people and races are fitter lay behind the "science" of eugenics, which culminated in the holocaust and the Nazis' horrific experiments with genetics. Some people still believe that humans are innately aggressive, competitive, and selfish; yet the evidence is overwhelmingly against them. Most of the world's population follow religions that specifically frown on aggression and selfishness. Even non-religious people usually condemn them, and altruistic acts – a firefighter plunging into a burning building to rescue someone – are common.

How do we acquire a set of morals? We only develop the power to make moral judgments as we mature. Moral sense comes, it seems, with increasing ability to reason; from social learning from parents and teachers; with the development of emotion, especially empathy; and from biological needs and drives. American psychologist Lawrence Kohlberg sees us as going through three moral stages. He used as an example the dilemma of a man whose dying wife can be saved only by a drug sold by a profiteering chemist. He cannot raise the money, so decides to steal the drug. In the first stage (7–10 years old), when we simply try to avoid punishment, we would worry only about his getting caught or his wife dying. In the second, when moral conventions guide our judgments (10–16 years), we might reason that it is bad to steal or good to take care of a dying wife. In the third stage, adults judge for themselves and might worry about the social consequences of stealing or the balance between preserving life and upholding the law.

EVERY ACT AN ANIMAL ACT?

A creature that helps others – especially at risk to itself – will, in strictly Darwinian terms, reduce its chances of surviving and producing offspring. So we would expect evolution to rule out altruism. And yet the animal world shows plenty of examples of altruistic behavior that would be seen as "moral" in humans.

A rabbit endangers itself by stopping to thump its feet to warn other rabbits of the approach of a predator. Songbirds sing to warn other birds of danger – even though by singing they draw attention to themselves. And despite the fact that they will produce no offspring, sterile worker bees spend all their lives sustaining the queen.

One answer to this apparent paradox is the idea of the selfish gene – a theory which has also been applied to human actions – that altruism is really selfishness at the genetic level. A mother fox will risk her life to save her cubs because her genes will survive in them.

Another theory is that of "delayed reciprocal altruism," the idea that one good turn deserves another. Young female baboons, for instance, groom dominant adult females who may, later, become powerful allies; young males groom less dominant females who, in turn, let them mate.

WHAT IS A SIN?

Every society and every culture has its own system of morals, from the Ten Commandments handed down from God to Moses in Judeo-Christianity to the pronouncements of the Islamic prophet Muhammad. Sometimes these are given extra weight by religious dogma or are enforced by the power of the law – yet even societies which have no organized religion or legal system almost always have an underlying guiding morality, as is evident from the table on the right.

What is remarkable about moral systems is not only how diverse they are, but also how much they have in common.

	Anglican	Catholic	Judaism	Islam	Hinduism	Buddhism	Secular
Blasphemy	Sin	Sin	Wrong	Sin	Not a sin	Not applicable	Acceptable
Non-observance of religious events	Sin	Sin	Wrong	Sin	Not a sin	Not harmful	Normal
Murder	Sin	Sin	Wrong	Sin	Sin	Harmful	Wrong
Adultery	Sin	Sin	Wrong	Sin	Sin	Harmful	Wrong
Theft	Sin	Sin	Wrong	Sin	Sin	Harmful	Wrong
Lying	Sin	Sin	Wrong	Sin	Sin	Harmful	Wrong
Premarital sex	Sin	Sin	Wrong	Sin	Sin	Not harmful	Normal
Homosexual practices	Sin	Sin	Wrong	Sin	Sin	Not harmful	Normal
Divorce	Not a sin	Sin	Permitted	Not a sin	Sin	Not harmful	Acceptable
Masturbation	Sin	Sin	Wrong	Not mentioned	Sin	Not harmful	Normal
Suicide	Not a sin	Sin	Wrong	Sin	Sin	Harmful	Acceptable
Cruelty to animals	Not a sin	Not a sin	Permitted	Not a sin	Sin	Harmful	Wrong

Feeling guilt is one of the most notable instances of the moral mind at work. Most of us feel uneasy if we do something we perceive – or know – to be wrong. In Japan, such feelings have been exploited at notorious accident locations by installing plastic policemen; these artificial enforcers of the law have proved effective in slowing down the traffic and preventing accidents.

Indeed, a feeling of guilt can be so powerful that it lasts a lifetime, and a misdeed committed in early childhood can haunt us for many years. We are not always aware of our guilt, yet it rises to the surface in unconscious ways, affecting our behavior, like Shakespeare's sleepwalking Lady Macbeth rubbing her hands and crying, "Out damned spot!" as she tries to rid herself of the guilt of killing Duncan.

See also

STATES OF MIND
▶ Talking cures 164/165

▶ Motivation 180/181

▶ Individual minds 184/185

SURVEYING THE MIND
▶ The working mind 16/17

▶ Comparing brains 18/19

BUILDING THE BRAIN
▶ The brain plan 68/69

INPUTS AND OUTPUTS
▶ Survival sense 122/123

FAR HORIZONS
▶ The evolving mind 126/127

▶ Learning to be human 144/145

▶ Milestones of the mind 146/147

▶ Thinking together 150/151

Individual minds

Each person's unique mind leads to unique thoughts and actions, but do individuals choose what they do?

Should a gangster's son turn to lawlessness, people say "like father, like son," as if crime was an inevitable consequence of his parentage. We accept that our physical make-up comes from our parents. Could it be that behavior does, too? One branch of psychology – sociobiology – says so. It argues that everything we do, everything we are, is a product of evolution and our genetic inheritance.

Many behaviorist psychologists, on the other hand, argue that what we do is determined by our past experiences; our lives, they say, are completely "programmed" by the pattern of rewards and punishments we encounter. What both the behaviorists and the sociobiologists have in common, though, is that they suggest that our every action is predetermined and inevitable. This is not as outrageous as it sounds. All classical science works on the premise that there is a cause for every event and that one thing inevitably leads to another.

In fact, most psychologists must work on this assumption, too – that there is a reason for everything we do or feel. But if everything we do or feel is caused inevitably by something else, do we have any real freedom of choice? Is there such a thing as free will? The problem with this idea is that it runs counter to our own personal experience and to the way society is run. We all feel as if we are making free choices (within certain constraints) all the time, and our legal system is based on the assumption that we are each responsible for our actions.

To deal with this apparent paradox, many psychologists adopt something similar to the idea of "soft determinism." They believe that for psychology as a science, determinism works best, but since there is more to life than science, free will works best in relation to subjective, everyday experience. Essentially, soft determinists say everything has a cause, but they recognize free will as a cause, just like any "external" one. They argue that we act freely when we are neither constrained nor coerced. Clearly, a prisoner is constrained from doing what he or she wants to do; or a gun held to the head may coerce someone into doing something he or she does not want to. But the idea allows constraints and coercion

Leonardo da Vinci's self-portrait is just one of his many remarkable works of art. Were these and his brilliant scientific studies due to his own genetic allocation or were they a result of his inevitable experiences and memories of his environment? Many would say it was Leonardo's unique genius; deterministic psychologists would counter that he could not help but do what he did.

to work much more subtly and deeply than this. We might be constrained by a childhood fear, for instance.

Still other psychologists maintain that to contrast free will with determinism is misleading. They argue that, as we are material beings in a world where everything has a cause, our actions must also be based in causality. But since the causes are so numerous, they are effectively indeterminate, and so we have freedom of choice. It is, they say, our biology that makes us free because our biology has made us into creatures who are always creating our own mental and material environments.

OUT OF THE SILENT DARKNESS

Born a normal baby, Helen Keller (1880–1968) suffered a fever at 19 months which left her deaf and blind, and soon afterward she became unable to speak. With only her senses of smell, taste, and touch remaining, the child was, in effect, cut off from the world. When Helen was about 6 years old, her parents sought help with her education, and a remarkable 20-year-old woman – Anne Sullivan – came to teach her. Anne, who had herself been blind at one time but had had her sight partially restored, was a graduate of a school for the blind. Anne "spoke" to Helen by spelling words on her palm, teaching her the names of objects which Helen could feel. Later Anne helped Helen to learn to speak by allowing Helen to feel the vibrations in her throat as she spoke.

Helen proved such an adept pupil that by the age of 10 she could read Braille, and at 24 she graduated from college with the constant help of Anne, who accompanied her to classes where she spelled out the lectures on Helen's hand. Soon after graduating, Helen wrote her biography, later becoming a famous author. She dedicated her life to helping the blind and deaf and went on worldwide lecture tours to promote their education. Everyone who met her, including writer Mark Twain, was enthralled by her vivacity, warmth, and intelligence. Stories like this, of human beings who live remarkable lives despite severe handicaps, persuade many people to believe that there is more to being human than genetic programming. There is little more striking evidence of the willpower of the individual human mind than Helen Keller's story.

Helen Keller's unique circumstances generated a unique individual. Unable to hear or see, she was nevertheless able to communicate and make an extraordinary contribution to the world around her.

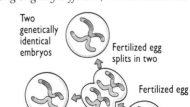

Identical twins are as alike as two humans can be – their genes are identical. Yet despite physical similarities, they have unique minds. The genes they share can be expressed in divergent ways, and subtle differences in experiences will give each one unique capabilities and memories, and thus a unique way of responding to events.

The coming together of an egg and a sperm creates a genetically unique fertilized egg, which develops into a genetically unique person. No other human egg has had the same genes – nor will it have in the future – unless it splits in two, which happens occasionally. In that case, two genetically identical eggs form, giving rise to two genetically identical twins. But almost from the moment the two identical eggs appear, environmental influences act upon them, and the embryos, later babies, and then adults develop along slightly different, individual lines.

Two genetically identical embryos

Fertilized egg splits in two

Fertilized egg

Eggs

Mother

Father

Sperm

See also

STATES OF MIND
▶ The unique self
156/157

▶ Being aware
158/159

▶ Motivation
180/181

▶ The moral mind
182/183

BUILDING THE BRAIN
▶ The brain plan
68/69

▶ Linking up
74/75

FAR HORIZONS
▶ The evolving mind
126/127

▶ The infinite store?
130/131

▶ Active memory
132/133

▶ Having ideas
140/141

Bibliography

Agur, Anne and Ming Lee *Grant's Atlas of Anatomy* Williams and Wilkins, Baltimore, Maryland and London, 9th ed., 1991

Blakemore, Colin et al. *The Mind Machine* BBC Books, London, 1988

Carlson, Neil R. *Physiology of Behavior* Allyn and Bacon, Inc., Boston and London, 3rd ed., 1986

Carola, Robert et al. *Human Anatomy and Physiology* McGraw-Hill, Inc., New York, 1992

Comer, Ronald J. *Abnormal Psychology* W.H. Freeman and Company, New York, 1992

Cooper, Jack R. et al. *The Biochemical Basis of Neuropharmacology* Oxford University Press, Oxford and New York, 6th ed., 1991

Crick, Francis *The Astonishing Hypothesis: The Scientific Search for the Soul* Touchstone Books, London and New York, 1994

Dawkins, Richard *The Selfish Gene* Penguin, London, 1986

Dennet, Daniel C. *Consciousness Explained* Penguin, New York, 1993

Diamond, Marian C. et al. *The Human Brain Coloring Book* Barnes & Noble Books, New York and London, 1985

England, Marjorie A. and Jennifer Wakely *A Colour Atlas of the Brain & Spinal Cord* Wolfe Publishing Ltd., London, 1991

FitzGerald, M.J.T. *Neuroanatomy Basic and Applied* Baillière Tindall, London, 1985

Giese, Arthur C. *Cell Physiology* W.B. Saunders Company, Philadelphia, Pennsylvania, 1979

Glees, Paul *The Human Brain* Cambridge University Press, Cambridge and New York, 1988

Gould, Stephen Jay *Wonderful Life* Hutchinson Radius, London, 1989
——*Ever Since Darwin* Penguin, London, 1991

Graham, Robert B. *Physiological Psychology* Wadsworth Publishing Company, Belmont, California, 1990

Greenfield, Susan G. *Journey to the Centers of the Mind: Toward a Science of Consciousness* W.H. Freeman and Company, New York, 1995

Gregory, R.L. *Eye and Brain: The Psychology of Seeing* Oxford University Press, Oxford, 1990
——*Oxford Companion to the Mind* Oxford University Press, Oxford, 1987

Guyton, Arthur C. *Textbook of Medical Physiology* W.B. Saunders Company, Philadelphia, Pennsylvania, 6th ed., 1981

Huxley, Aldous *The Doors of Perception* Perennial Library, New York, 1954

Jones, Steve *The Language of the Genes* Flamingo, London, 1994

Kandel, Eric R. et al. *Principles of Neural Science* Prentice-Hall International Inc., London and New Jersey, 1991

Koestler, Arthur *The Ghost in the Machine* Macmillan, New York, 1967

Kuffler, Stephen W. et al. *From Neuron to Brain* Sinauer Associates Inc. Publishers, Sunderland, Massachusetts, 2nd ed., 1984

Levitan, Irwin B. and Leonard K. Kaczmarek *The Neuron: Cell and Molecular Biology* Oxford University Press, Oxford and New York, 1991

Lyons, A. and R.J. Petrucelli, II *Medicine – An Illustrated History* Abradale Press/Harry N. Abrams Inc. Publishers, New York, 1987

Nieuwenhuys, Rudolf *Chemoarchitecture of the Brain* Springer-Verlag, New York, 1985

Ornstein, Robert *Psychology* Harcourt Brace Jovanovich, New York, 1988
——*The Evolution of Consciousness: The Origins of the Way We Think* Touchstone, New York, 1992

Parker, Steve *Eyewitness Science: Human Body* Dorling Kindersley, London, 1993
——*Eyewitness Science: Medicine* Dorling Kindersley, London 1995

Penrose, Roger *The Emperor's New Mind* Oxford University Press, Oxford, 1989

Pinchot, Roy B. (ed.) *The Brain* Torstar Books, New York, 1984

Silverstone, Trevor and Paul Turner *Drug Treatment in Psychiatry* Routledge, London and New York, 5th ed., 1995

Smith, Anthony *The Mind* Hodder & Stoughton, London, 1984

Smith, Jillyn *Sense and Sensibilities* Wiley Science Editions, New York and Chichester, 1989

Snyder, Solomon *Drugs on the Brain* Scientific American Inc., New York, 1986

Steinberg, Robert *In Search of the Human Mind* Harcourt Brice, Orlando, Florida, 1995

Steur, Faye B. *The Psychological Development of Children* Brooks/Cole Publishing Company, Pacific Grove, California, 1994

Strickberger, Monroe W. *Evolution* Jones and Bartlett Publishers, Boston, 1990

Williams, Peter L. et al. (eds.) *Gray's Anatomy* Churchill Livingstone, Edinburgh, 38th ed., 1995

Wills, Christopher *The Runaway Brain* Flamingo, London, 1993

Winfree, Arthur T. *The Timing of Biological Clocks* Scientific American Inc., New York, 1987

Zeki, Semir *A Vision of the Brain* Blackwell Scientific Publications, London and Boston, 1993

Suggested Periodicals and Journals
New Scientist IPC Magazines Ltd., London
Scientific American Scientific American, Inc., New York

IN SHORT – WHAT THE LETTERS MEAN

ADP	adenosine diphosphate
ANS	autonomic nervous system
ATP	adenosine triphosphate
CAM	cell adhesion molecule
CAT	computerized axial tomography
CNS	central nervous system
CSF	cerebrospinal fluid
ECT	electroconvulsive therapy
EEG	electroencephalogram

5-HT	5-hydroxytryptamine (serotonin)
fMRI	functional magnetic resonance imaging
GABA	gamma-aminobutyric acid
LGN	lateral geniculate nucleus
MEG	magnetoencephalography
MRI	magnetic resonance imaging
PET	positron emission tomography
PNS	peripheral nervous system
PVS	persistent vegetative state

MEASURE FOR MEASURE

In the metric system a millimeter (mm) is a thousandth of a meter ($\frac{1}{25}$ inch); a micrometer, or micron (μm), is a millionth of a meter ($\frac{1}{25,000}$ inch); and a nanometer (nm) is a billionth of a meter ($\frac{1}{25,000,000}$ inch). A millivolt (mv) is a thousandth of a volt (V); a microvolt is a millionth of a volt.

In this book, a billion is 1,000 million, or 10^9; a trillion is a million million, or 10^{12}.

Index

Page numbers in *italic* refer to captions and box text.

A

a capella singing *99*
acetic acid *61*
acetylcholine
    breaking down *61, 176*
    circadian rhythms in production *121*
    discovery *52, 60*
    function *37*
acetylcholinesterase *61*
actin *53*
action potentials 40, *40, 46, 47, 52*
    generation 48, *48–49*
    and polarization 56, *56*
    structural basis 50–51, *50–51*
acupuncture *105*
adenosine *179*
adenosine diphosphate *see* ADP
adenosine triphosphate *see* ATP
ADH *118, 119*
ADP *66, 67*
aging 76, *76–77*
agricultural revolution *127*
alcohol 77, *178, 178, 179*
algorithms *23*
alpha waves *27*
altruism 182, *182*
Alzheimer's disease 77, *176, 176–77*
amacrine cells *45, 91*
Ames Room *95*
amniotic cavity *68*
amoeba *12,* 40
amphetamines *178, 179*
ampullar nerve *113*
amygdala *34, 36, 154*
amygdaloid complex *108*
analgesics *104*
aneurysm *65*
angular gyrus *33, 139*
animal studies 18, *18–19*
anions *46, 47*
anorexia nervosa *174*
anoxia *76*

ANS *see* autonomic nervous system
anterior commissure *36, 108*
anterior corticospinal tract *83*
anterior olfactory nucleus *36, 108*
anterior spinocerebellar tract *83*
antidepressant drugs *166–67, 168–69*
antidiuretic hormone *see* ADH
antipsychotic drugs *166*
ants *151*
anxiety *37, 163, 172*
anxiety disorder *172, 172–73*
apes
    modern *126*
    terrestrial *126*
Aplysia 62
ApoE protein *176*
apoptosis *77*
a posteriori knowledge *14*
a priori knowledge *14*
arachnoid membrane *43*
Archaeopteryx 13
Aristotle *14, 15*
arousal *155*
artificial intelligence (AI) 22, *23*
Ascidian 13
aspartate *60*
association 21, *130, 155*
association areas *35*
association cortex *86*
association fibers *31*
association tracts *31*
astrocytes *43, 64*
astrology *156*
atoms *46*
ATP *60, 66, 67*
auditory cortex *97, 114, 139*
auditory ossicles *96, 96*
auditory perception 98, *98–99*
auditory radiations *97*
auditory reflex *96*
australopithecines *126*

Australopithecus afarensis 13
Australopithecus robustus 126
autism *144*
autonomic nervous system (ANS) *81, 116, 116–17, 118*
    development *71*
    parasympathetic *81*
    sympathetic *81*
autoradiography *27*
autoreceptors *173*
axonal flow *61*
axon hillock *48, 48, 49, 54, 56*
axons 41, 44, *45*
    development 74, *74–75*
    growth cone *75*
    injury *64*
    length *103*
    myelinated *50*
    signals along 50–51, *50–51*
axon terminal *53, 60*

B

babies, newborn 142, *142–43*
Babinski reflex *82*
backbone *82*
Baez, Joan *172*
balance 87, 112, *112–13*
barbiturates *167, 179*
Barnum statements *156*
barosensors *116, 117*
basal ganglia *86*
bats, navigation *19*
battery *47*
bees 74, *151*
Beethoven, Ludwig van *99*
behavior, abnormal *162, 162*
behaviorism 16, *17, 18, 184*
behavior therapy *164, 165, 175*
benzodiazepines *167, 173*
bereavement *169*
beta waves *27*
biological psychology *17*

biorhythms 120, *120–21*
bipolar cells *90, 91*
bipolar disorder (manic-depression *163, 167, 168*
birds
    brain *13*
    flight *19*
    vision *90*
birth 70, *133*
Black Death *127*
blastocyst *68*
blindsight 25, *92*
blind spot *88, 94*
blood–brain barrier 42, *43, 66, 77*
blood flow to organs *66, 67*
blood pressure *119*
blushing *119*
body language *134–35*
body temperature 35, *119, 181*
bonobos, sign language *135*
Braille, Louis *101*
Braille system *101*
brain
    anatomy 30, *30–31, 32, 32–33*
    blood flow *66, 67*
    blood supply *29*
    and central nervous system 80, *81*
    comparison with computers 22, *23*
    damage 24, *24–25,* 64, *64–65, 77, 138, 159*
    energy use 66, *66–67*
    evolution 12, *12–13*
    fetal development 70, *70–71*
    function map 34, *34–35*
    imaging 28, *28–29*
    investigating the function of 26, *26–27*
    maps 30, *30–31, 32, 32–33*
    neurotransmitter map 36, *36–37*
    protection 42, *42–43*
    size and weight *40*
    terminology *32*

brain stem *13, 116, 116, 117*
breathing control *117*
Broca, Paul *15, 24*
Broca's area 24, *25, 137, 139*
Brodmann's areas *35*
Bufo periglenes 148
bulimia nervosa *174*
buspirone *173*

C

Cade, John *167*
caffeine *179*
calcarine sulcus *33*
calcium ions *77*
cambium *72*
capsaicin *106*
cardio-acceleratory area (CAA) *116*
cardio-inhibitory area (CIA) *116*
cardioregulatory center *116*
cardiovascular center *116*
catfish *106*
cations *46, 47*
cats, balance *112*
CAT scans 28, *28, 29*
Cattell, Raymond *157*
caudate nucleus *31, 34, 36*
CCK *60*
cell adhesion molecules (CAMs) *72*
cell differentiation *68*
cell groups A11, A13, A14 *36*
cell membrane *46, 47*
cells 40
    structure *41*
    suicide genes *75, 77*
    see also individual cell types
cell-substrate adhesion processes *72*
cellular respiration *66*
central executive *131*
central motor command/ program *86, 87*
central nervous system (CNS) *13, 81*
    development *71*

see also brain; spinal cord
central sulcus *33*
centriole *41*
cerebellar peduncles *33*
cerebellum *12, 13*
    anatomy *30, 33*
    and balance *112*
    damage to *25*
    development *70*
    function *35*
    and memory *131*
    and movement *86, 87*
cerebral arteriography *29*
cerebral cortex *19, 82*
    anatomy *31, 33*
    and balance *112*
    development 71, *72*
    and emotions 154, *154*
    function 35, *138, 138–39*
    and senses *115*
    size and intelligence *115*
cerebral embolism *65*
cerebral ganglion *12*
cerebral hemispheres *15*
    anatomy *30, 31, 33*
    development *70*
    evolution *13*
cerebral thrombosis *65*
cerebrocerebellum *87*
cerebrospinal fluid (CSF) *15, 82*
    protective function 42, *42, 43*
cerebrum *12, 13*
    anatomy *30, 31, 33*
    development *70*
cervical ganglia *120*
chemosenses *106, 106, 108*
chemosensors 102, *106–7, 107, 116, 117*
childhood development
    brain 74, *74–75*
    mental 146, *146–47*
chili peppers *106*
chimpanzees, sign language *135*
chlordiazepoxide *167, 173*
chloride ions *46*
chlorpromazine *166*
cholines *37, 61*

choroid 88
choroid plexus 42
chromosomes 148
ciliary muscles 88
cingulate gyrus 33
cingulate sulcus 33
circadian rhythms 120, 120–21
cirrhosis of the liver 178
citric acid cycle 67
CNS see central nervous system
cocaine 178, 179
cochlea 96, 96, 97
cochlear nerve 36, 96, 97
codeine 179
coercion 184
cognitive death 159
cognitive labeling 154
cognitive needs 181
cognitive psychology 17
cognitive therapy 165
collateral sulcus 33
Collembola 12
color vision 90
    defects 90
coma 159
communication 127, 134, 134–35, 150
communities 150–51
computerized axial tomography see CAT
computers, comparison with brain 22, 23
concrete operational stage of development 146
conditioned responses 128, 129
conditioning 128, 129, 131
cone cells
    mode of action 89
    position 88
    transmission of information from 90, 90–91
    and visual perception 94, 94, 95
consciousness 14, 158, 158–59
constrictor pupillae muscles 93
convergent analytical skills 140
cornea 88, 88

corpus callosum 27
    anatomy 31
    cutting 15, 139
    function 34, 138
crack 178
cranial nerves 80, 81, 82
    anatomy 30
    and autonomic nervous system 117
    damage to 25
    and pain perception 102
    and sense of smell 108
    and sense of touch 100
    and taste 107
cranium 42
creativity 140
crista 113
cuneiform symbols 136
cuneus 33
curiosity 180
cyclic guanosine monophosphate (cGMP) 89
cytoplasm 41

D

dark current 89
Dartmuthia 12
da Vinci, Leonardo 184
deafness 99
death, cognitive 159
deep pyramidal cells 109
delta waves 27
dementia 176, 176–77
dendrites 41, 44, 45
    development 74, 74–75
dendritic spine 45, 53
dentate gyrus 33
depolarization 51, 56, 56
depressant drugs 179
depression 61, 163, 168, 168–69
    endogenous 169
    exogenous 169
dermis 100, 102
Descartes, René 14
determinism 184
diazepam 167, 173
diet 77
divergent analytical skills 140
DNA 41, 45
dogs, sense of smell 18, 108
dolphins 70

dopamine 59, 60
    effects of stimulants on levels 179
    function 36
    receptors 24, 166, 171
    in schizophrenia 170, 171
dorsal diagonal nucleus 36
dorsal fornix 33
dorsal longitudinal fasciculus 36
dorsal nerve roots 100
dorsal tegmental nucleus 36
dorsal vagal nerve nucleus 36
double-bind situations 171
downers 179
Down's syndrome 142, 176
dreams 95, 160, 160–61
    interpretation 165
dressing up 158
drives 155, 166, 180, 181
drug abuse 77, 178, 178–79
drug dependence 178
drug therapy in mental illness 166–67, 166–67
drug tolerance 178
drug withdrawal symptoms 178, 179
dura mater 43

E

eagles, vision 90
eating disorders 174
echolocation 98–99
Ecstasy 178, 179
ECT 166, 166
ectoderm 69
Edinger Westphal nuclei 93
EEG 26, 27
    in epilepsy 56
    in sleep 120–21, 160
ego 16
Einstein, Albert 40
electroconvulsive therapy see ECT
electroencephalography see EEG
electrons 46
electron transport system 67

elephants 75
embryonic development 68, 68–69, 70, 72, 72–73
embryonic disk 68, 69
emotional development 144–45, 144–45
emotions 154, 154–55
endocytosis 61
endoderm 69
endoplasmic reticulum 41, 53
endorphins 60, 104, 106, 169, 179
energy, potential 47
enkephalins 37, 60, 104
entorhinal area 108
environmental deprivation 121
ependymal cells 43
ependymal layer 73
epidermis 100, 102
epidural hemorrhage 65
epilepsy 27, 56, 61
    grand mal 56
    petit mal 56
epinephrine 122
Erisistratus 15
estrogen 148
ethology 17, 18, 18–19
eugenics 182
evolution 126, 126–27
    brain 12, 12–13
excitatory postsynaptic potential (EPSP) 56, 56, 57, 58
exercise
    and depression 169
    and relaxation 173
exocytosis 53
extensor thrust reflex 112
exteroceptors, cutaneous 100
eye 32
    insect 89
    structure and function 88, 88–89
Eysenck, Hans 157

F

F1 particles 67
facial nerve 80, 107
faculties 14
FAD3 gene 176
false sago palm 77
fertilization 68
fetal alcohol syndrome 142

fetal position 70
fetishism 174
fetus
    development 70, 70–71, 72, 72–73, 142
    developmental abnormalities 142
    REM sleep 160
fight or flight response 117, 119, 122–23, 172
filum terminale 82
fish
    brain 12
    pheromones 111
5-HT see serotonin
5-HT1A autoreceptor 173
fixed-action patterns 18
flavors 107
flicker fusion 94
flies 89, 106
flocculonodular lobe 35
flooding 175
fluoxetine 61, 168, 169
folic acid 142
fornix 32, 33, 34, 36
fovea 88, 90, 94
free association 165
free nerve endings 100, 101, 102
free radicals 76
free will 184
Freud, Sigmund 17, 160, 164, 165
frontal lobes
    damage to 24, 25
    function 35
functional magnetic resonance imaging (fMRI) 28–29, 29

G

GABA 59
    in anxiety 173
    discovery 60
    effect of marijuana on 179
    function 37, 56
Gage, Phineas 24
Galen 15, 157
Gall, Franz 15, 16
gambling 129
gamma-aminobutyric acid see GABA
ganglion 80
    cells 88, 90, 91, 92
Gardner, Howard 141

gastric ulcer 123
gate theory 102
gender differences 148–49, 148–49
genes 68
German measles see rubella
Gestalt psychology 17
Gestalt therapy 165
glial cells 43, 64
globus pallidus 31, 34, 36
glossopharyngeal nerve 107
glucose 66, 67
glutamate 59, 60, 63
glycine 59, 60, 63
glycolysis 67
Goethe, Johann Wolfgang von 14
golden toad 148
Golgi, Camillo 40
Golgi apparatus 40, 41
Golgi tendon organs 85
G-protein linked receptors 58, 59
G-proteins 59
grammar 136–37
grasp reflex 143
grief 169
group efforts 150
groups, belonging to 150–51
group therapy 164
guilt 183
gustation 106–7, 106–7
gyri 32, 33

H

habenula 36
habenulo-interpeduncular tract 36
habituation 62
hair, protective function 42, 43
hair cells 113
Haller, Albrecht von 15
hallucinations 110, 161
hallucinogenic drugs 179
harmony 99
head 13
hearing 96–99, 96–97, 98–99
    loss 99
heart, blood output 67
heart disease and stress 172

heart rate 119
  control of 116
Helmholtz, Hermann von 15
heroin 178, 179
Herophilus 15
Hertz (Hz) 96
hippocampus
  destruction 27
  function 34, 36
  glutamate receptors 63
  and memory 131
  in schizophrenia 170
  and sense of smell 108
  terminology 32
Hippocrates 14, 157
histamine 60, 102
homeostasis 118, 118–19
hominids 13, 126
Homo erectus 126, 134, 150
H. habilis 126
H. sapiens 122, 126, 150
H. s. neanderthalensis 126
H. s. sapiens 126
horizontal cells 91
hormones 118–19, 148, 180
Hughes, Holly 140
human genome 68
humanistic therapies 165
hunger 119, 180, 181
hydrogen peroxide 76
Hylonomus 12
hyperpolarization 56, 56
hypnosis 175
hypochondria 163
hypoglycemia 67
hypothalamus 21
  anatomy 31
  and drives 180, 181
  effects of antipsychotic drugs on 166
  and emotions 154, 154
  endorphin production 104
  function 34, 34, 36
  and homeostasis 118, 118–19
  links with brain stem centers 116
  and taste 107
hypoxia 76
Hypsognathus 12
hysteria 163

I
imaging of the brain 28, 28–29
implosion therapy 164, 175
imprinting 18, 145
incus 96, 96
infant development
  brain 74, 74–75
  emotional 144–45, 144–45
  mental 146, 146–47
infant prodigies 147
inferior colliculus 97, 98
inferior frontal gyrus 33
inferior frontal sulcus 33
inferior temporal gyrus 33
inflammatory response 102
influenza 142
infrasound 99
infundibular nucleus 36
inhibitory postsynaptic potential (IPSP) 56, 56, 57
injury 102
insects
  brain 12
  eye 89
instinct 180, 181
intelligence 140, 141
  bodily kinesthetic 141
  interpersonal 141
  intrapersonal 141
  linguistic 141
  logical-mathematical 141
  multiple component theory 141
  musical 141
  personal 141
  spatial 141
intelligence quotient tests see IQ tests
intention 35
intermediary cells 88
internal capsule 31
interpeduncular nucleus 36
interpersonal relationships 144
intestinal blood flow 67
introspection 16
ions 46, 46, 47, 48
iproniazid 166–67
IQ tests 140, 141
iris 88, 88

J, K
Jaynes, Julian 16
jellyfish 12
jet lag 121
Kanizsa illusion 94
Kant, Immanuel 14
Keller, Helen 185
Klüver-Bucy syndrome 154
knowledge
  a posteriori 14
  a priori 14
Kohlberg, Lawrence 182
Korsakoff's syndrome 178
Krause endings 101
Krebs cycle see citric acid cycle

L
language 127, 134, 136–37, 136–37, 139
  learning 137
  pidgin 137
  written 136
Largactyl 166
Lascaux cave paintings 126
Lashley, Karl 130
lateral cerebral sulcus 33
lateral corticospinal tract 83
lateral geniculate nuclei (LGNs) 32, 92, 92, 93, 114, 115
lateral lemniscal nuclei 36
lateral occipital sulcus 33
lateral occipital temporal gyrus 33
lateral parabrachial nucleus 36
lateral spinothalamic tract 83, 102, 103
L-dopa 171
learned helplessness 168
Leeuwenhoek, Antoni van 14
lemniscal tracts 83
lens 88, 88
lentiform nucleus 32, 34
libido 154, 180
Librium 167, 173
life events 123, 172
limbic system
  effects of antipsychotic drugs on 166
  input to hypothalamus 118
  and sense of smell 108, 110
  and sensory information 114
  and taste 107
lingual gyrus 33
lion 148
lipid peroxidation 76
lithium 167
Locke, John 14
locus coeruleus 36, 37
Loewi, Otto 52
logic 14
loners 151
Lorenz, Konrad 17, 145
love 154
lysergic acid diethylamide (LSD) 179
lysosomes 41

M
macroglia 43
macrophages 43, 77
magic mushrooms 179
magnetic resonance imaging see MRI
magnetoencephalography 28–29
malleus 96, 96
mammalian brain 13
mammillary body 33, 34
mammillothalamic tract 36
manic-depression see bipolar disorder
marijuana 179
Maslow, Abraham 181
MDMA see Ecstacy
mechanoreceptors 101
medial dorsal nucleus 108
medial frontal gyrus 33
medial geniculate nucleus 97, 114
medial longitudinal fasciculus 31, 96
medial occipital temporal gyrus 33
medial temporal gyrus 33
medial temporal lobe 177
medial temporal sulcus 33
medulla 34, 85, 103, 116

Megazostrodon 13
Meissner's endings 100–1
melatonin 120
memory 130
  declarative 131
  episodic 130, 131
  false 132
  feats of 133
  flashbulb 133
  jogging 133
  long-term 131
  non-declarative 130, 131
  seat of 132
  semantic 130, 131
  sensory 20, 21
  short-term/working 20, 131
  storage 132
  types 131
meninges 42, 42, 43, 65, 82
menopause 148
mental development 146, 146–47
mental illness 162, 162–63
  physical therapies 166–67, 166–67
  psychotherapy 164–65, 164–65
meperidine 179
meridians 105
Merkel endings 101
mescaline 179
mesoderm 69
metarhodopsin II 89
methadone 179
methionine 104
Meynert's nucleus 36
microfilaments 45
microglial cells 43, 77
microtubules 41, 45, 53
midbrain 34, 36, 85, 103
Mill, James 14
mind–body dualism 14
mitochondria 41, 45, 53
mnemonics 133
monism 14
monoamine oxidase inhibitors (MOAIs) 168
monoamine re-uptake inhibitors (MARIs) 168
morals 182, 182–83
Moro reflex 143
morphine 104, 179
moths 111
motivation 180, 180–81

motor cortex 84, 85, 86, 87
  left 139
  primary 35
motor decussation 31, 85
motor homunculus 84
movement 84, 84–85
  ballistic 86, 86
  sensory-guided 86, 86
Mozart, Wolfgang Amadeus 147
MPTP 24
MRI 28–9, 29
Müller, Johannes 15
multicelled organisms 12
multiple personality disorder 156
multiple sclerosis (MS) 51
muscles
  blood flow 67
  control of 84, 84–85
muscle spindles 85
musical ability 138
myelin 43, 45, 50–51, 50, 51, 74

N
narcotic drugs 179
necrosis 77
need hierarchy 181
neglect 25
Neher, Erwin 49
neocortex 72, 73
nerve cells see neurons
nerve growth factor (NGF) 75
nerve plexus 81
nerve signals see action potentials
neural crest 69, 71
neural groove 69
neural networks 22, 23
neural plate 69
neural tube 69, 71, 82
neurites 44, 44, 45
  development 74, 74–75
  see also axons; dendrites
neuroblasts 72, 72, 73, 75
neurofilaments 45
neuroglia 42, 43
neurology 40
neurons
  activation 48, 48–49
  alterations in strength of connections 62, 62–63

autorhythmic 48
bipolar 44
in brain injury 64
development 71, 72,
   72–73
electrical properties
   46, 46–47
electronic versions 23
evolution 12, 12
excitable cell
   membrane 46
function 40, 41
glomerular 45
Golgi 44
investigating the
   function of 26, 26–27
motor/efferent 15, 80,
   85
pioneer 75
postsynaptic 60
presynaptic 60
Purkinje 45
pyramidal 44
recording activity 27
sensory/afferent 15,
   80
stellate 44
structure 41, 44, 44–45
target 75
tonotopic arrangement
   97
unipolar 44
neurosis 163
neurotransmitter-gated
   ion channels (NGICs)
   58, 58, 59
neurotransmitters 62, 63
   action of psychoactive
      drugs on 179
   action of psychotropic
      drugs on 166–67,
      166, 167
   in Alzheimer's disease
      176
   circadian rhythms in
      levels 120, 120–21
   convergence 60
   divergence 60
   excitatory 36, 56, 60
   function 48, 52, 52–53
   inhibitory 36, 56, 60
   map of 36, 36–37
   receptors see receptors,
      neurotransmitter
   recycling 43, 60–61
   research on 60–61, 60

in schizophrenia 170,
   171
neurotubules 61
newborn babies 142,
   142–43
nicotine 178, 179
nicotinic receptor 59
nitrous oxide 60
NMDA receptors 63
nociceptors 101, 102
node of Ranvier 45, 50, 51
non-verbal
   communication 134–35
norepinephrine 59, 60
   circadian rhythms in
      production 121
   during dreaming 161
   effects of
      antidepressants on
         levels 167
   effects of ECT on
      levels 166
   effects of stimulants
      on levels 179
   function 37
notochord 69
nucleus
   atom 46
   cell 41, 44, 45
   nervous system 80
nucleus interstitialis
   striae terminalis 36
nucleus solitarius 36, 107

O
obesity 181
obsession 174
obsessive-compulsive
   disorder 163, 172, 174
occipital lobes 92
occipital temporal sulcus
   33
OCEAN 157
odor molecules
   (odorants) 108, 109, 110
olfaction see smell, sense
   of
olfactory bulb 12, 108,
   108, 109, 110
olfactory cortex 109
olfactory epithelium 108
olfactory glomeruli 108,
   109
olfactory nerves 108
olfactory receptor cells
   108, 108, 109, 110

olfactory tract 33, 108,
   109
olfactory tubercle 36, 108
oligodendrocytes 43
olivary nuclei 36, 97, 98
ommatidia 89
operant conditioning 164
operants 128
operational stage of
   development 146
ophthalmometer 15
ophthalmoscope 15
opiates 179
opioids 179
opium 179
opsin 89
optic chiasma 32, 92, 93,
   120
optic nerve 32, 88, 90, 92,
   94, 115
optic radiations 92, 93
optic tracts 32, 92, 93
organelles 41, 44
organ of Corti 96, 97
osmoreceptors 119
otolith organs 112, 113
oval window 96, 96
ovaries 148
oxygen 66

P
Pacinian endings 100–1
pain
   pathways 102, 102–3
   perception 35, 100,
      104, 104–5
   relief 104, 105
   and survival 123
   threshold 105
   types of 104
palate 106
paleocortex 108
panic attacks 172
parabrachial nuclei 36
paracentral lobule 33
parahippocampal gyrus
   33
paraquat poisoning 24
pararterminal gyrus 33
paraventricular nucleus
   36
paravertebral ganglia 117
parietal lobe damage 24,
   25
parietal occipital sulcus
   33

Parkinson's disease 24,
   61, 77, 176
passive spread 48
patch clamping 27, 49
Pavlov, Ivan 17, 128, 129
PCP (phencyclidine) 179
Penfield, Wilder 132
penguin, emperor 148
periaqueductal gray area
   (PAG) 104
periglomerular cells 109
peripheral nervous
   system (PNS) 43, 81, 100
   and pain pathways 102
persistent vegetative
   state (PVS) 159
personality 156, 157
   disorders 156
   types 156, 157
perspective 95
PET 28–29, 28, 140, 177
phenothiazines 166, 171
pheromones 111
phobias 128, 163, 164,
   172, 174–75, 174–75
phonological loop 130,
   131
phosphodiesterase 89
phosphorus 66, 67
phrenology 15, 16
Piaget, Jean 146
pia mater 43
pictograms 136
pidgin language 137
pike 110
pineal gland 14, 120
pituitary 34
   and drives 180
   endorphin production
      104
   and homeostasis 118,
      118–19
placenta 70
planning 35
plasma membrane 41,
   44
Plato 14
play 75, 144
PNS see peripheral
   nervous system
pons 36
   and control of
      breathing 117
   and control of
      movement 85
   function 34

and pain perception
   103
   vasomotor center 116
population 127
positron emission
   tomography see PET
postcentral gyrus 33
postcentral sulcus 33
posterior spinocerebellar
   tract 83
postsynaptic potential
   (PSP) 48, 48–49, 53,
   58, 62
   excitatory 56, 56, 57,
      58
   inhibitory 56, 56, 57
post-traumatic stress
   syndrome 133, 172
posture 87
potassium ions 46, 47
potential energy 47
precentral gyrus 33
precentral sulcus 33
precuneus 33
prefrontal cortex 170
prefrontal lobe damage
   25
prefrontal lobotomy 166
premotor cortex 35
pre-neurons 72, 72, 73,
   75
preoperational stage of
   development 146
preoptic nucleus 36
presynaptic excitation 55
presynaptic facilitation
   62
presynaptic inhibition 55
pretecta 93
primary reinforcers 128
pro-hormones 118
projection fibers 31
proprioception 83, 112
prosopagnosia 95
protein synthesis during
   sleep 161
Prozac 61, 168, 169
psilocybin 179
psychoactive drugs 179
psychoanalysis 17, 160,
   164–65, 165
psychology 16, 17
psychosis 163
psychosurgery 166, 166
psychotherapy 164–65,
   164–65

psychotropic drugs
   166–67, 166–67
puberty 146
pupil 88, 88, 93
putamen 31, 34, 36
pyramidal decussation
   31, 85
pyramidal tracts 83
pyriform cortex 108

R
radial glioblasts 72, 73
Ramapithecus 13
raphé nuclei 36, 104
Ravel, Maurice 138
Reagan, Ronald 176
reasoning 14, 147
receptors
   neurotransmitter 36,
      53, 58, 58–59
   in brain injury 64
   effects of drugs on
      167
   hippocampus 63
   in schizophrenia 171
   toxin uptake 24
   skin 100, 100–1
   tongue 107
reflexes 82, 96, 112, 128,
   142, 143
   primitive 143
reinforcers 128
relaxation 173
REM sleep 120–21, 160,
   160–61
reproductive cycle 119
reptiles, brain 12
respiration
   cellular 66
   control of 117
respiratory center 116, 117
respiratory rate 119
resting potential 46, 47
reticular formation 118
reticulospinal tracts 83
retina 32, 92, 94
   function 90, 90–91
   neural cells 89
   structure 88, 88–89
retinal 89
retro-rubral area 36
rhodopsin 89
ribosome 60
rod cells
   mode of action 89
   position 88

structure 88
transmission of information from 90, 90–91
and visual perception 94, 94, 95
Roentgen, Wilhelm 28
rooting reflex 143
Rorschach, Hermann 157
Rorschach inkblot test 157
rubella 142
rubrospinal tract 83
Ruffini endings 101

S
saccule 112, 113
Sakmann, Bert 49
salmon, migration 110
saltatory conduction 51
scala media 97
schizophrenia 163, 170, 170–71
and dopamine 36
and flu during pregnancy 142
hearing voices 16
and neurotransmitters 61
therapy 166, 166
Schwann cells 43
sclera 88
sea squirt 13
secondary auditory cortex 97
secondary olfactory cortex 108
secondary reinforcers 128
secondary sexual characteristics 148
second messenger systems 59
sedatives 166, 167
seizures 56
selective serotonin re-uptake inhibitors (SSRIs) 168
self, sense of 16
self-actualization 181
self-awareness 158, 158
selfish gene 182
semicircular canals 113
sense organs 12
sensorimotor stage of development 146
sensory homunculus 100
septal nucleus 34

septum 32, 36
septum pellucidum 33, 34
serotonin (5-HT) 59, 60, 61
in anxiety 173
effects of antidepressants on levels 167, 168
effects of ECT on levels 166
effects of stimulants on levels 179
function 36
sex differences 148, 148–49
sex drive 154, 180
sexual activity 119
Siffre, Michel 121
single-celled organisms 12, 40
6-OH dopamine 24
Skinner, B.F. 17, 128, 129
skin receptors 100, 100–1
skull 42
sleep 120, 120–21, 160, 160–61
deprivation 161
smell, sense of 108–11, 106, 115
in dogs 18
effect on emotions and memory 110, 110–11
structural basis 108, 108
smell maps 108–9
smoking 142
snakes, heat radiation sense 19
social insects 151
social phobias 174–75
social readjustment rating scale (SRRS) 173
society 150–51
sociobiology 184
sodium ions 46, 47
soft determinism 184
somatosensory cortex 35, 100, 100, 103, 139
somatosensory receptors 107
somatosensory system 100
somatostatin 60
somites 69
sonar 98–99
soul 158
sound
perception of 96, 96–97
pitch 96
waves 96

spatial summation 49
speech 134, 136–37, 136–37
speed 179
spiders, brain 19
spina bifida 142
spinal column 82
spinal cord 12, 13
anatomy 30
and central nervous system 80, 81
development 69, 70
injury 64
and pain pathways 102, 103
structure and function 82, 82–83
spinal nerve roots 81, 100
spinal nerves 30, 36, 81, 83
spinal tracts 83
spinocerebellum 87
spinoreticulothalamic tract 83, 102, 103
sponges 12
stage fright 172
stapes 96, 96
stellate cells 109
stepping reflex 143
stereotyped behavior 174
stimulant drugs 179
stimuli 129
stress 122–23, 122–23, 172, 173
stress disorders 172
stressors 123, 172
stria medullaris 36
stria terminalis 36
stroke 65
subarachnoid space 42, 43
substance P 60
substantia nigra 24, 36, 170, 176
subthalamic nucleus 36
suicide 163
suicide genes 75, 77
sulci 32, 33
Sullivan, Anne 185
superficial pyramidal cells 109
superior central nucleus 36
superior colliculus 36, 92, 93, 96, 98
superior frontal gyrus 33
superior frontal sulcus 33
superior temporal gyrus 33

superior temporal sulcus 33
speech 134, 136–37, 136–37
superstitions 128, 128, 129
suprachiasmatic nuclei 36, 119, 120
supraoptic nucleus 36
survival of the fittest 181, 182, 182
survival mechanisms 122–23, 122–23
sweating 119
synesthesia 114
synapse (synaptic gap) 45, 56
action of psychotropic drugs at 166, 167
axoaxonic 55
axodendritic 54
axosomatic 54
axospinodendritic 54, 55
axospinosomatic 54
in brain injury 64
dendrodendritic 55
excitatory 55, 57
formation 74, 74–75
inhibitory 55, 57
passage of nerve signal across 48, 48, 52, 52–53
types 54, 54–55
systematic desensitization 164, 175

T
TA 165
Tabanis 89
tacrine 176
taste 106–7, 106–7
taste buds 106
taste centers 107
tectorial membrane 97
temporal lobes 92
damage to 25
function 35
and memory 132
temporal summation 49
Ten Commandments 183
tendons 84
termites 18, 151
testosterone 148
thalamus 20
anatomy 31, 32
and emotions 154
and movement 86

and pain perception 103
and sense of smell 108
structure and function 114, 114–15
and taste 107
thermosensors 101, 119
theta waves 27
thinking 14, 14–15, 20, 20–21
thirst 119
Thorazine 166
tibial nerve 80
time zones 121
Tinbergen, Niko 17
tobacco 178, 179
tongue 106
touch 35, 100–1, 100–1, 102
toxins 77
tranquillizers 179
transactional analysis 165
transcendental meditation 173
transference 165
transoccipital sulcus 33
tree rings 72
trepanning 14, 15
tricyclic antidepressants 168
trigeminal nerve 80, 100, 106
trigeminal nerve nucleus 36
Tripp, Peter 161
Turner's syndrome 148
twins, identical 185
tympanic membranes 96
tyramine 168

U
ultrasonic sound 98
umbilical cord 70
uncus 33
utricle 112, 113

V
vagus nerve 107, 117
Valium 167, 173
vasomotor center 116
ventral diagonal nucleus 36
ventral nerve cord 12
ventral tegmental area 36, 170

ventricles 31, 42, 43
vermis 35, 87
vertebrae 82
vertebral canal 82
vertebrate brain 12, 13
vervet monkeys, communication 135
Vesalius, Andreas 15
vesicles 41
synaptic 52, 53
vestibular nerve 97
vestibular nucleus 36
vestibulocerebellum 87
vestibulocochlear nerve 97
vestibulospinal tract 83
vision 88–95, 90–91, 92–93
visual association areas 35
visual cortex 20, 35, 92, 93, 114, 115, 139
damage to 25
visual field 92
visual information pathways 90
visual perception 94, 94–95
visual–spatial scratch pad 131
visual startle reflex 93
visual tracking reflex 93
vitalism 14
vitreous humor 88
voltage-gated calcium channels 53
voltage-gated ion channels 49, 50, 50, 51, 57

W, X, Y
walking 143
walking reflex 143
Watson, John B. 16
Wernicke's area 137, 139
damage to 25
Wertheimer, Max 17
whales 98–99
use of sound 98–99
Wiltshire, Stephen 144
Wundt, Wilhelm 16
X chromosomes 148
X-rays 28, 28, 29
Y chromosomes 148
yellow spot 88, 90, 94
yolk sac 68

Acknowledgments

l = left; *r* = right; *c* = center; *t* = top; *b* = bottom.
JB = John Barlow

Picture credits
2*t* Mehau Kulyk/Science Photo Library; 2*b* JB; 10*l* Michael W. Davidson/Science Photo Library; 10*tr* Science & Society Picture Library; 10*br* JB; 11 Warren Anatomical Museum, Harvard Medical School, Boston; 12/13 Animals Unlimited; 14 Science & Society Picture Library; 16 JB; 18/19 Alain Compost/ Bruce Coleman; 19, 20 JB; 20/21 Dan Bosler/Tony Stone Images; 21*t&br* JB; 21*bl* Warren Faioley/ Oxford Scientific Films; 22*t* Manfred Kage/Science Photo Library; 22*b* Michael W. Davidson/Science Photo Library; 23*tl&b* JB; 23*tr* US Department of Energy/Science Photo Library; 24 Warren Anatomical Museum, Harvard Medical School, Boston; 26*t* Guy's Hospital anatomy department; 26*b* Mike Powell/Allsport; 28 Tim Beddow/Science Photo Library; 28/29*t* Mehau Kulyk/Science Photo Library; 28/29*b* The National Hospital for Neurology and Neurosurgery; 38*t* JB; 38*b* Dr. Pietro De Camilli, Yale University School of Medicine, and Dr. Michela Matteoli, University of Milan; 40/41 Steve Hopkin/Planet Earth Pictures; 43 Guy's Hospital Anatomy Department; 46/47 JB; 48 Bruce Forster/Tony Stone Images; 49 Max-Planck-Institut für Medizinische Forschung; 50/51 Lori Adamski Peek/Tony Stone Images; 52 Don Fawcett/ Science Photo Library; 54/55 John David Begg; 56 Science Photo Library; 56/57 Leo Mason/The Image Bank; 58, 60 JB; 62 G.I. Bernard/Oxford Scientific Films; 62/63 Duncan Wherrett/Tony Stone Images; 64/65 Litsios/Frank Spooner Pictures; 65 Mehau Kulyk/Science Photo Library; 66/67 JB; 68 John Lawlor/Tony Stone Images; 69, 72 JB; 73 Mary E. Hatten; 74/75*t* Dr. Pietro De Camilli, Yale University School of Medicine, and Dr. Michela Matteoli, University of Milan; 74/75*b* Harvard University Press, graphic from the postnatal development of the human cerebral cortex by Jesse Leroy Conel vol. 1, 1939, 1959; 76 The Kobal Collection; 78, 79, 80 JB; 82*t* Mel Lindstrom; Tony Stone Images; 82*b*, 84, 85 JB; 86 Bolcina/Frank Spooner Pictures; 89 Claude Nuridsany and Marie Perennou/Science Photo Library; 90 Institute of Opthalmology; 90/91 Stephen Dalton/NHPA; 92/93 Images Colour Library; 94/95*t* Emile Luider/ Rapho/Network; 94/95*b* Maxim Ford; 96 JB; 98 James D. Watt/Planet Earth Pictures; 99*t* Eve Arnold/Magnum Photos; 99*b*, 100 JB; 101*t* Terry Vine/Tony Stone Images; 101*b*, 103 JB; 104 Allsport;

106*b* Anthony Bannister/NHPA; 106/107*t* JB; 108 Frank Spooner Pictures; 110/111 Roger Hutchings/ Network; 111, 112/113 JB; 115 Kim Westerskov/ Tony Stone Images; 116 JB; 117 Alon Reininger/ Colorific!; 118 F. Henry/Rea/Katz; 120/121 Paul Lowe/Network; 121 Michel Siffre; 122 JB; 122/123*t* James King-Holmes/Science Photo Library; 122/123*b* Museum of London; 124*l&tr* JB; 125*l* Michael Holford; 125*r* JB; 126/127 L.D. Gordon/The Image Bank; 128 JB; 128/129 Range/Bettmann; 129 Nostalgia Amusements; 130 JB; 132 William Feindel, Wilder Penfield Archive, Montreal Neurological Institute; 132/133 Petit Format/Bubbles; 133 JB; 134/135*t* Anna Clopet/Colorific!; 134/135*b* JB; 136 Michael Holford; 136/137 Wellcome Dept. of Cognitive Neurology/Science Photo Library; 138 Lou Jones/The Image Bank; 138/139 Dr. Colin Chumbley/Science Photo Library; 140 E. Ferrorelli/ Colorific!; 141 JB; 142 Sterling K. Claren, Prof. of Pediatric Dept., University of Washington School of Medicine; 142/143 JB; 144/145 Kunsthistorisches Museum, Vienna/The Bridgeman Art Library; 145*t* Range/Bettmann/UPI; 145*b*, 146/147 JB; 148 Michael Fogden/Oxford Scientific Films; 149 Penny Gentieu/Tony Stone Images; 150 Alex Webb/ Magnum Photos; 150/151 Robert Harding Picture Library; 152*tl* Image courtesy of Peter Ramm, Imaging Research Inc.; 152*tr* James Stevenson/ Science Photo Library; 152*bl* JB; 152*bc* Rorschach H., Rorschach-Test © Verlag Hans Huber AG, Bern, Switzerland, 1921, 1948, 1994; 152*br* Neil Fletcher; 153 JB; 154/155 Werner Bokelberg/The Image Bank; 155 JB; 156/157 C. Poulet/Frank Spooner Pictures; 157 Rorschach H., Rorschach-Test © Verlag Hans Huber AG, Bern, Switzerland, 1921, 1948, 1994; 158/159 JB; 160 James Stevenson/Science Photo Library; 160/161 Image courtesy of Peter Ramm, Imaging Research Inc.; 162 Antonio Ribeiro/Frank Spooner Pictures; 162/163 Eric Bouvet/Frank Spooner Pictures; 164 JB; 164/165 Paul Grendon/ Select; 165 Freud Museum, London/© Sigmund Freud; 166/167, 168 JB; 168/169 Don McCullin/ Magnum Photos; 169 Bruce Ayres/Tony Stone Images; 170/171 JB; 172/173 David Redfern/ Redferns; 174 Ariel van Straten; 175 Neil Fletcher; 176 Tim Beddow/Science Photo Library; 176/177 Range/Reuter/Bettman; 177 Tim Beddow/Science Photo Library; 178 Eugene Richards/Magnum Photos; 178/179 JB; 180*t* Matt Meadows/Science Photo Library; 180*b* JB; 181*t* Nigel Cattlin/Holt Studios International; 181*b* JB; 182 The Kobal Collection; 183 Douglas Dickins; 184 Biblioteca

Reale, Torino/Scala; 185*t* Range/Bettmann; 185*b* Michael Nichols/Magnum Photos

Illustration credits
David Ashby 40/41; John Barlow 84/85, 100/101; Richard Bonson 40/41, 42/43, 44/45, 72/73, 76/77; Bill Donahoe 24/25, 52/53, 58/59, 60/61, 92/93, 166/167; Andrew Farmer 14/15, 16/17, 18/19, 26/27, 74/75, 88/89, 92/93, 94/95, 110/111, 118/119, 126/127, 140/141, 142/143, 146/147, 158/159, 172/173, 174/175, 178/179; Chris Forsey 12/13, 80/81, 82/83, 86/87, 106/107, 112/113, 130/131, 162/163, 170/171; Gary Hinks 68/69, 70/71, 114/115; Sally Launder 96/97, 108/109, 118/119; Mainline Design 20/21, 24/25, 46/47, 48/49, 50/51, 54/55, 56/57, 62/63, 66/67, 84/85, 90/91, 104/105, 116/117, 120/121, 148/149, 156/157, 160/161, 168/169, 172/173, 176/177, 182/183, 184/185; Mike Saunders 64/65, 84/85, 100/101, 102/103, 150/151; Technical Art Services 136/137; Richard Tibbets 30/31, 32/33, 34/35, 36/37, 138/139, 170/171; Zhang Tongyun 104/105; Mark Watkinson 46/47; Stephen Wiltshire 144/145

Marshall Editions would like to thank the following:
Authors: Jerome Burne 24–29, 126–51, 156–61, 176–77
John Farndon 12–23, 154–55, 162–75, 178–85
Steve Parker 40–77, 80–123

The Independent (June 8, 1995); Olympus Sport for man's track suit bottom; Quicks Archery for bow and arrow; Chappell of Bond Street, London, for Beethoven's bust; Freed of London for woman's leotard; David Mellor for pressure cooker, cruet, casserole and cooking utensils; Chess & Bridge for chess board and pieces; BMW for car keys; Body Active for dumbells; Reject Shop for cushions. The Family Flynn. John Barlow for retouching and imaging. Jo Neild for cooking the casserole. Dawn Lane for make-up.

Dr. P.W. Atkins, University Lecturer in Physical Chemistry and Fellow in Chemistry, Lincoln College, Oxford. George Bridgeman at Guy's Hospital Anatomy Department. Dr. John Coleman and Mr. Andrew Slater, Oxford University Phonetics Department. Dr. Huw Dorkins, Senior Research Fellow and Tutor in Medicine, St. Peter's College, Oxford. Dr. Andrew King, Senior Research Fellow, University Laboratory of Physiology, Oxford University. Dr. John Morris, University Lecturer in Human Anatomy and Wellcome-Franks Tutor, St. Hugh's College, Oxford. Dr. J. Stein, Lecturer in Physiology, Oxford University Medical School, and Fellow of Magdalen College, Oxford.